D0959591

FRUIT

Rodale's Home Gardening Library™

FRUIT

edited by Anne M. Halpin

 Rodale Press, Emmaus, Pennsylvania

Printed in the United States of America on recycled paper containing a high percentage of de-inked fiber (black and white pages only).

Book design by Marcia Lee Dobbs and Julie Golden
Illustrations by Kathi Ember and Jack Crane
Photography credits: Mark Kane: photo 11; Rodale Press Photography Department: photos 1, 2, 3, 7, 8, 9; Jack Ruttle: photos 4, 5, 6; W. Atlee Burpee Co.: photo 10.

Library of Congress Cataloging-in-Publication Data

Fruit.

 (Rodale's home gardening library)
  1. Fruit-culture. 2. Organic gardening. 3. Fruit.
I. Halpin, Anne Moyer. II. Series.
SB357.24.F78   1988   634   87-23309
ISBN 0-87857-736-X   paperback

2   4   6   8   10   9   7   5   3   1   paperback

Contents

1

Planning the Fruit Garden

Nowadays, we can purchase fruits from all over the world at the local supermarket. But there are special rewards in growing our own, and savoring the juicy sweetness of tree-ripened, chemical-free fruit. An astonishing variety of fruits will grow within the confines of the United States and Canada. Fruits that are grown for commercial use — picked when immature and shipped long distances — cannot compare in flavor to homegrown varieties picked at the peak of ripeness. And many lesser-known fruits can offer a whole new range of taste treats to the gardener.

Along with their edible qualities, fruits are a welcome addition to the backyard landscape, too. Fruit trees provide welcome shade on hot summer days, and a sturdy place for children to climb (especially apple trees, with their strong branches and wide crotches). When positioned properly, groups of trees can serve as a windbreak. And trees decorate the yard with their fragrant flowers in spring and colorful foliage in fall. Fruiting shrubs and vines are useful, too. They can be trained on trellises or along fences to serve as privacy screens, or to hide a shed or garage from view.

Deciding What to Grow

Chances are that no matter where you live, the combination of sunshine, temperature, soil, moisture level, and wind that make up the growing environment will support several kinds of delicious fruit. The key to success is understanding the environmental conditions you have to offer, and choosing the fruits that will grow best under those conditions. If you want to be successful at growing fruit, you will have to do a little research and experimentation to

1

determine what fruits you will be able to grow in your geographical area and what conditions you need to provide for their best growth. You will also need to assess carefully your family's preferences, your storage capacity, and the amount and kind of growing space you have available. In this chapter we will be giving you some general guidelines to follow. You can get more specific information on conditions in your area from your local extension agent of the U.S. Department of Agriculture (USDA). Other good sources of information include local nurserymen, your newspaper's garden editor, growers, amateur orchardists, and members of orcharding societies. It's a good idea to talk to more than one person and get a variety of opinions. Let's go on then, and examine the factors that you will need to investigate when determining what kinds of fruits you can grow.

Climate

The first and most important consideration in deciding what to grow is climate. You will be able to accommodate only those fruit species and varieties that can stand the most extreme temperatures of the year in your geographical area. The familiar temperate-zone fruits, which are all deciduous, are widely grown and have been highly developed horticulturally. Many of these fruits have been extensively bred, and many varieties have been developed that are adapted to a range of growing conditions; there are varieties with increased tolerance of heat and cold, and others with greater resistance to insects and disease. Dwarf varieties are available which yield sooner and produce greater yields per acre than standard-sized fruit trees. There are well over 600 varieties of apples alone grown in the United States. The adaptability of temperate fruit trees (at least of trees that have been grafted, as most available varieties are) to climate is determined primarily by the rootstock, although the climatic preference of the scion wood must also be considered.

Plants from temperate areas can grow within a range of temperatures, but all of them require a cold period each year (the chilling factor) in order to bloom and set fruit. Generally speaking, temperatures of 45°F or lower over a period of at least 45 days will be a sufficient cold period. If temperatures are too cold for a prolonged period, blossom buds can be damaged or destroyed.

Apples are generally the hardiest of fruits—their buds can withstand temperatures down to −25°F. Least hardy of the temperate fruits are peaches and nectarines, which can sustain damage if temperatures fall below −12° to −15°F. Between these two extremes are (in order of decreasing hardiness) pears, European plums, sour cherries, sweet cherries, Japanese plums, and apricots.

When climatic requirements have been assessed, there remains the challenge of choosing not just kinds of fruits but specific cultivars that are suited to the winter climate in your area, and to the spring and summer growing conditions as well. Besides average winter temperatures, other characteristics of temperate fruits include whether there are sudden swings from warm to cold weather or strong winds that make damage from cold more likely even at usually acceptable temperatures, and whether summers are too long, hot, or dry for good results. Finally, you must determine whether the fruit's bloom date is so early as to risk damage from late spring frosts. This is one of the greatest risks faced by fruit growers, who must take into account not just wood hardiness but that of the flower buds as well.

Family Likes and Dislikes

Given the range of fruits it is possible to grow in your part of the country, the next step in planning the orchard or garden is to take a good look at what fruits your family likes, and how you will use the harvest. Along with the number of people in your family, how you are going to use the fruit will guide you in determining how much of each type to plant. If your family likes apples, of course, you'll want to eat some of the fruit fresh. But will you also want to bake pies or make cider or applesauce? You need to know which apple varieties are best suited to your purpose, and how much of a yield to expect from your plants, bushes, or trees.

Choosing the Site

Gardeners with little space and no choice of sites need to choose fruits that will tolerate the environmental conditions they have to offer. But for gardeners with a large property, too much emphasis can't be placed on the importance of careful site selection for the orchard or berry patch. After you're familiar with the general environmental conditions of your area, you'll need to carefully assess your property to determine the best fruits for your location and the best location for your fruits.

First, keep in mind that growing fruit means a long-term commitment of space. Most vegetables complete their life cycle in a single growing season. Most fruit trees and small fruit plants, however, are perennials which live from three to forty or more years. Programmed for a long life, fruits direct much of their strength and energy toward building a durable root system that will sustain them through extremely cold or dry seasons. Because of these extensive roots and considerable size at maturity, most fruits are difficult to move once they're established. The decision as to which ones to plant and where to plant them requires both thought and foresight. It's important to be aware of the conditions you have, but don't let them scare you away from growing fruit. If you choose your crops and varieties carefully, you can grow fruit just about anywhere.

Environmental Conditions on Your Property

In general, fruits grow best with full sun and good (but not excessive) air movement to minimize frost damage and the spread of fungal diseases. In warm, humid areas, exposure to full sun is extremely important; a southern and eastern exposure will help the sun dry morning dew quickly, making the environment less favorable for bacteria and disease-causing organisms. The soil should be deep and well drained, so that no surface water remains in puddles after a prolonged rainy spell. At the same time, the soil must retain moisture long enough for roots to absorb it, and must be of moderate fertility. Excessively rich soils, especially those abundant in nitrogen, will produce lots of weak shoot growth at the expense of the fruit.

Most fruits need a steady supply of moisture in order to grow well. If your soil tends to dry out, mulching can help to hold in moisture. Another technique, especially useful with young trees, is to leave the soil around the trunk slightly depressed, to catch every bit of rainfall and prevent runoff.

If your soil is a heavy, slow-draining clay, or has hardpan (a hard layer below the surface that resists water and root penetration) or bedrock less than 3 feet below the surface, you can install drainage tiles before you plant the trees or bushes. Soils with a higher percentage of sand and organic matter will have better drainage to begin with and shouldn't need any modification below the surface.

Soil pH should also be considered, and it can be modified somewhat. Overly acid soil can be limed, and alkaline soils can be conditioned with peat moss, oak leaves, or cottonseed meal. Most fruits grow comfortably within a slightly acid range of pH values, although some are quite specific in their requirements. Watermelons and blueberries, for example, need a very acid soil with a pH

If your soil tends to dry out, leave a small depression around newly planted trees to serve as a reservoir for rainwater and to catch runoff moisture.

between 4 and 6 to do well. (See Chapter 5 for information on the pH needs of individual fruits.)

It's important to consider air drainage, too, when determining what types of fruits your property can support and where you will put them. Keep fruits out of low-lying areas like valleys and river-bottoms where cold air can settle and form frost pockets. Cold air, like water, seeks low spots, and you may lose crops if you plant in a cold pocket. Instead, put your trees and bushes on a rise where cold air will drain away. If you have a hilly piece of property, do not plant your fruits closer than 50 feet to the floor of a low-lying area. Conversely, keep them away from the crest of the rise where the trees could be exposed to temperature extremes and drying winds. The ideal setting would be on the slopes of high, rolling hills, perpendicular to prevailing winds and away from cold areas. If you must plant in a low-lying spot, you can lessen the danger of a cold pocket by aligning the rows of trees in the same direction as the air flows.

You must also consider the direction of the slopes. There are advantages and disadvantages to both the north and south sides of hills. South-facing slopes receive more sunshine than the north and are generally warmer—good conditions for plant growth. But plants grown on southern slopes often flower earlier, which can make them susceptible to damage from late spring frosts. Less hardy fruits like apricots, peaches, and sweet cherries should be planted on the north-facing sides of hills so that they will flower later and avoid the frosts. On the other hand, north-facing slopes are cooler and will delay the spring bud break and bloom. Trees planted in cool places may exhibit a tendency for the bark to crack in the winter unless painted white.

Another consideration in locating fruit plantings, for gardeners who live in windy areas, is the presence of a suitable windbreak to protect the fruit trees from prevailing winds in spring and summer. A windbreak will protect an area on the downwind side that is five times the height of the windbreak. Fruit trees should not be planted closer than one windbreak-height away from the windbreak. In urban areas, shade trees and nearby buildings may give enough wind protection.

A final climatic factor to keep in mind is that large bodies of water, such as lakes, will help to moderate the climate of nearby areas. This effect is most pronounced downwind of the water, where humidity will be increased and winter temperatures not as severe. The humid air will stay cooler longer in spring, meaning that trees will tend to bloom a bit later than normal and are not likely to be damaged by late spring frosts. Areas which experience frequent foggy conditions will also experience delayed bloom. However, during the summer the presence of fog may encourage mildew and other diseases in fruits, and can reduce yields.

Size and Space Needs

The size and shape of your property directly affect the number and kind of fruits that you grow. Obviously, a large piece of property will accommodate a number of types of fruits, but a small property will limit your choice. Keeping in mind that some of the fruits will need cross-pollination, which makes it necessary to plant more than one fruit of that type, your next consideration after climate, family preferences, and soil and site would be the amount of space needed for each bush or tree to grow to maturity.

Dwarf vs. Standard Varieties

In recent years, dwarf and semidwarf trees of the most popular temperate tree fruits have been on the market in increasing numbers. Dwarf varieties are a boon to commercial growers and home gardeners with limited space—they produce full-sized fruit on a tree that is much smaller than the normal size. They are often more productive per unit of space, too. Dwarf trees are usually more expensive than standard-sized ones, but they begin to bear at

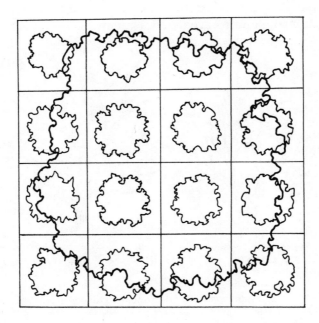

Sixteen dwarf trees can be grown on the same amount of ground that a single standard tree would normally occupy.

an earlier age and are easier to prune and harvest. A tree is considered to be a dwarf when it is less than half the size of the standard tree at maturity. A semidwarf tree is one that reaches 50 to 85 percent of the size of a standard tree.

Smaller trees can be created in a number of ways: by budding or grafting standard varieties onto dwarfing rootstocks, by tree-training practices such as espalier or bonsai, and by breeding and selection. Some of the popular dwarf varieties, such as the NORTH STAR cherry, are genetic dwarfs, which are kept small throughout their lifetimes by their own genetic make-up.

Another kind of tree dwarfed by mutation is the "spur-type" tree which has been selected from a mutant or "sport" branch growing on a standard tree. These sport branches are full of short, fruit-bearing spurs rather than long and leafy branches. Generally, spur-type trees are smaller, grow more slowly, and produce larger crops sooner in their lives than standard varieties. The spur trees are also sometimes grafted or budded onto dwarfing rootstocks, in which case they are referred to as "double-dwarf." In nursery catalogs, spur-type trees are listed as such.

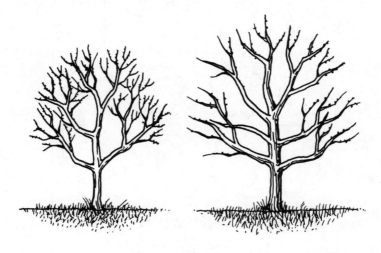

A spur-type tree (left) has fewer lateral branches and more fruit-bearing spurs than a standard tree (right). It also bears fruit more heavily while being somewhat smaller than a standard tree.

Most of the fruit trees sold by nurseries are actually parts of two different trees: a rootstock which is selected to give the tree a firm, strong anchor in the ground, and a scion, the top part, which is selected for the type of fruit it bears. The most commonly employed method of producing dwarf or semidwarf trees is by grafting scions onto rootstocks that reduce their growth. For example, quince rootstocks are used for dwarfing scions of standard-sized pears. Scions are chosen for characteristics like the quality of their fruit, market availability, and resistance to disease. A dwarfing rootstock creates a smaller tree by impeding the flow of nutrients

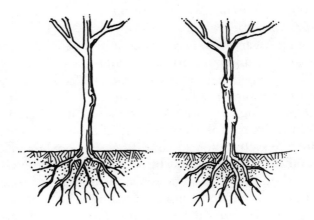

Many fruit varieties have inferior root systems despite their excellent fruit characteristics. To overcome this problem, a scion of a good fruiting variety is grafted onto a rootstock of a strong-rooted variety, as shown on the left. Sometimes an interstock is grafted between the scion and the rootstock (shown on right) to avoid incompatibility between the stock and the scion or to make use of its growth-controlling properties. Graft-unions are recognized by bulges on the trunk.

between the scion and the roots. The leaves of the scion produce more food than the rootstock can absorb and transport. As a result, carbohydrates build up in the scion and encourage earlier production of flowers and fruit.

Another way of grafting to produce dwarf trees is a two-stage process. In the first year, a dwarfing "interstock" is budded or grafted onto a rootstock that suits the particular soil and climate. Then the selected scion is budded or grafted onto the interstock the following year. This is done to improve the compatibility between scion and rootstock, or to create a truly dwarfed tree on roots that do not have true dwarfing capability. Such three-part trees are called "double-worked." Double-working is viewed by some nurserymen as the direction of the future for fruit trees, because it allows more opportunity to combine dwarfing, hardiness, and disease-resistant characteristics to produce trees ideally suited for the conditions provided in North American gardens.

The right cultural practices can also help to keep trees small. It is important not to overfeed trees, particularly with nitrogen, which would produce weak, rapid shoot growth that is susceptible to pests, diseases, and injury. Regular, moderate pruning also helps keep growth in check, and careful tip pinching and pruning in summer can help, too.

Although dwarf trees have come a long way in recent years, there are still problems, and work continues. One problem is incompatibility — between scions and rootstocks, and between rootstocks and American growing conditions.

When planning your orchard, whether you will grow dwarf or standard-sized trees, it is important to choose varieties and rootstocks that are well suited to conditions in your areas and your particular site. The trees must be able to tolerate the climate, of course, but they should also be resistant to the diseases that are common in your area. Some mail-order nurseries graft their varieties onto a number of different rootstocks matched to various geographic regions. A reputable local nurseryman or your local extension agent should be able to advise you about the best fruit varieties and rootstocks for your area.

Planning the Plot

When placing fruits relative to each other on your site, keep in mind the ultimate size of fruit trees and bushes, and situate them where they will receive the full sun that most of them need. Be sure to locate them far enough from the house and other structures so they will not eventually block the light or the view or become difficult to maintain. Trees close to the house can clog drains, crack foundations, or become damaged by the eventual digging of plumbers or utility crews. If you plan to plant fruit trees or bushes along property lines shared with neighbors, discuss your plans with them. They may not mind contending with occasional dropped fruit on their side if you offer to share some of the fruit with them.

You will probably want to locate standard-sized trees farther away from the house than your vegetable garden, for they won't need tending as often in one season, and the distance will be conducive to a better view of blossoms and ripening fruit. Dwarf trees can be managed around the fringes of vegetable gardens, and genetic dwarfs, especially peaches and nectarines, were developed with patios, front yards, and other landscaping situations in mind.

If you live where winters are cold, keep in mind the degree of hardiness of various species. The more tender fruits should be given comfortable settings — perhaps espaliered along a wall — and may need to be positioned so they can be sheltered from prevailing winds. (Likely sites might be where snow tends to melt first. Often this is a south-facing location backed on the north by a structure or hedge.) In very sunny but cold areas, a northeastern exposure may be advisable to prevent fruits from flowering too early, and to lessen the possibility of damage from sunscald.

Fruits must be spaced according to their size when mature, but that can vary considerably depending upon the cultivar, the soil, the care given, and the climate of the growing region. (For guidelines on spacing distances for individual fruits, see Chapter 5.) Where temperate fruit trees are concerned, dwarf forms often make the most sense for home gardens. A standard-sized tree may need as much as 20 to 40 feet; thus dwarf trees can convert the space needed for one standard tree into an orchard of several varieties chosen to ripen in succession to assure a minimal waste of fruit. Don't make your trees compete with one another for soil moisture and nutrients, though. Never plant a tree so that its dripline will overlap the dripline of another tree. This is especially easy to overlook when you're planting a small tree, and when planting in spring before neighboring trees are in full leaf.

Besides size, another variable also will influence the number (and kind) of fruits you can grow and the layout you choose. That factor is whether a fruit is self-fertile or whether two or more plants of different varieties are required for cross-pollination. If several varieties are indicated, position them in the same area — perhaps in alternate rows — to make visits by pollinators more likely.

Orchards

The arrangement of fruits in orchards should be determined by the lay of the land, the kinds and varieties of plants to be grown, and by soil management practices. Where annual rainfall is over 35 inches, permanent sod can be laid or a cover crop planted between the trees to add organic matter to the soil and help hold in moisture.

However, the soil inside the dripline of each tree should be cultivated or mulched. Sod and cover crops planted close to trees will compete with them for water and nutrients.

Trees can be laid out in a number of ways. In the often-used square system, which allows for easy access to standard-sized trees, a tree is placed in each corner of a square, the dimensions of which are determined by the spacing demanded by the fruit being planted when it matures. In the quincunx, or diagonal, setup, a fifth, filler tree that may be removed later if necessary is placed in the center of a square; this system yields more fruit per acre than the square and works well when different varieties are in alternate rows. In triangular, or hexagonal, placement, all trees are equidistant on a triangle, making possible the planting of more trees per acre than does the square system using the same spacing. In the hedgerow system, compact trees are planted side by side, and enough space is left between the double rows to allow for maintenance and harvest.

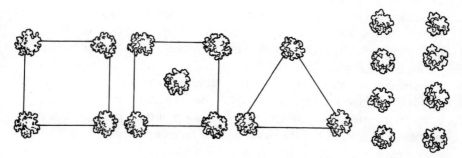

Here are four ways to lay out tree plantings (from left to right): trees are planted at each corner of an imaginary square in the square system; a fifth tree is added to the center of a square in the quincunx system; trees are placed equidistant on a triangle in the triangular system; and compact trees or bushes are planted side by side in the hedgerow system.

Garden Plots

Low-lying vining or bush fruits which don't require support may be grown in prepared plots either by themselves or in proximity to vegetables. Annual fruits such as melons thrive when two or three healthy plants grow in hills mounded about 4 inches above the surrounding soil. These mounds are spaced from 4 to 10 feet apart, depending on the varieties being grown. Permanent beds can be established for perennial berries.

Strawberries are sometimes grown in eight 10-inch-high hillrows, spaced 1 to 1½ feet apart — a planting arrangement often used for everbearing types. Unlike the matted row system, in which runners originating from the mother plants are left to fill in open spaces around the plants, strawberries grown in the hill system have all their runners removed, leaving only the mother plant at maturity. The hill system also can be used for upright raspberries and the bush forms of blackberries. Arching types are planted in rows in patches, as are blueberries, currants, and gooseberries, while trailing berry plants are trained to grow up or along trellises. Berries do well when mulched to keep weeds and annual plants from crowding them.

Trellises and Espalier

Fruit growers are increasingly appreciating the advantages of growing tree, bush, and vine fruits on trellises, or training them on wires as espalier. Trellising demands a lot of attention, but it is a great space-saver, and provides a way to grow fruits in places that might otherwise be too small to accommodate them. Yields can be higher, too. Some sort of trellis growing is mandatory for vine fruits such as grapes, and is recommended for plants with trailing canes, such as boysenberries and loganberries.

The profusion of flowers and fruits of superior quality which trellis growing produces in a greatly limited growing space is due to several factors which foster the growth of fruit-bearing spurs. These include generous feeding coupled with repeated pruning; the bending of branches toward the horizontal to slow growth (which is best done early in the growing season when new growth is limber); the optimal exposure of the center of the plant to sunlight; and the symmetrical shape of many espaliered forms, which facilitates an even flow of sap.

Although apples and pears are said to make the best espaliers, virtually all fruit trees and vining types of fruits are amenable to some form of trellising. Supremely elegant and often very practical, espaliered fruits are enjoying renewed popularity, and are a good choice for a narrow plot hemmed in by a pavement or driveway; for a trellis or fence marking property lines; or for a stunning specimen plant grown in a container on the patio.

In planning espaliered fruits, remember that white brick or concrete walls, especially if they face south, may cause plants to flower too early in spring or to become subject to overheating in summer. Painting tree trunks white will help keep them cooler and reduce the chance of sunscald. But in general, except in areas where winters are very severe and the heat radiated by a white wall would be helpful, a dark brick or stone wall facing east or west makes the best site for espaliered fruits. In placing these trained plants, also keep in mind any future necessity to move them in order to paint or maintain a wall.

Container Culture

Whether pruned to a natural shape, trained flat in espalier, or sculptured into a three-dimensional form such as a pyramid, fruits can be raised in containers. Since such plants — even when mature — can be grown in deep containers as small as 9 to 12 inches in diameter, a delightful variety of fruits can be enjoyed even in a small garden, yard, patio, or terrace. Fruit trees planted in containers will require more time and attention than trees grown in the ground. But for many people, container cultivation may be the only way to obtain fresh fruit.

The mobility of container fruits makes them a decorator's dream, and certainly containers are invaluable if you're renting or expect to move soon. A great advantage is that young trees can gain a head start toward maturity before you eventually put them into the ground in their permanent location. Indeed, many serious home orchardists have several contained trees tucked away in odd corners, waiting to fill in for a grounded tree lost to pests or diseases. Containers are also useful if you are growing a type of fruit or a particular variety not commonly grown in your area because you can observe its progress before committing it to a spot in the orchard.

In addition, containers make it possible to grow certain fruits beyond their normal geographic range. Northern gardeners with greenhouses or other suitable winter quarters can grow citrus, figs, or other subtropical or tropical fruits outdoors in summer and under cover in winter. Containers also make it possible to save your bounty from unseasonal frosts, damaging winds or hail, and hungry birds.

Of course, there is inevitably a certain price for such flexibility, for as British horticultural writer Alan F. Simmons has observed, "The more limited the space in which your fruit grows, the more work you will have to do." Potted fruits must have good drainage, topdressings of composted manure and other soil amendments as needed, more frequent watering than ground-planted fruits, periodic soil changes during dormant seasons, and either periodic root pruning or transplanting to a larger container. Container-grown fruits sometimes drop off the trees before they are ripe, possibly because of fluctuating moisture levels.

However, for adventurous gardeners or those with very little space, container culture is worth a try. It is currently an area of much experimentation, and new dwarf varieties that are better suited to container growing are being developed.

Landscaping with Fruits

The changing aspect of fruit plants and trees makes them decorative specimen plants in all seasons. In spring their fragile blossoms stir winter-dulled emotions; in summer their brightly colored bounty delights the eye and hand. In autumn the fiery foliage of deciduous trees warms the cooling air; and in winter their angled silhouettes offer beauty of a starker kind. This range of aesthetic effects can be enjoyed on every property, no matter how modest in size.

Potted dwarf fruits adorn patios or bright porches as little else can, while espaliered forms can glorify the otherwise uninteresting and unused wall of house, garage, or outbuilding. Small bush fruits such as currants can be used instead of more conventional flowering shrubs to frame a doorway or line a driveway or walkway. (Such borders also can be enhanced by strawberry plants or rows of dwarf fruits.)

Low-growing fruits are particularly handsome in front of stone walls. (The heat stored and gradually released by the masonry causes fruits to ripen faster and may even permit the successful culture of fruits usually raised farther south.) If the terrain slopes sharply, the effort of mowing grass and monotony of the usual groundcovers can be avoided by making terraces with bricks or railroad ties and creating a festive tiered strawberry patch.

Uninteresting fences take on a positive appeal when used as the backdrop for sculptured grapevines or neatly trained trailing bramble fruits. Another visually agreeable way to mark a property line is with a row of dwarf apples and/or pears planted 10 to 12 feet apart. Or instead of planting a privet hedge, how about one of blueberries, which will yield its sweet fruit in summer and flame scarlet in autumn; or a richly fruiting barrier of blackberries or upright raspberries.

Line your driveway with berry bushes to provide eye appeal as well as a profusion of juicy berries during the summer months.

Unattractive, idle corners of a yard can be brightened into color and productivity with a group of quinces, gooseberries, or brambles. Double rows of red raspberries, currants, elderberries—or perhaps grapes on a good-looking wooden arbor or raspberries trained on wires—also can form an attractive living screen to block a poor view or partition a yard.

For splendid ornamental effects based on contrasting forms and foliage, fruits can be interplanted. Black or purple raspberries, for example, make a graceful spreading shape that will do well in partial shade near dwarf fruit trees and complement their uprightness pleasantly. For other studies in contrast, bush fruits can be interplanted with other flowering shrubs — blueberries, for example, make handsome, culturally compatible companions for evergreens, since both favor an acid soil. Plantscapes in which fruits and flowers are combined can be especially picturesque, as when berries along a wooden fence form the background for a border of flowers, or daffodils or other early-flowering bulbs are planted in circular beds beneath dwarf fruits. Nearby flowers will also enhance pollination of fruits. A few blooming flowers tolerant of filtered light can be tucked under deep-rooted fruit trees pruned to an open center. (Be careful of planting too many flowers under your trees, though, lest they compete for the nutrients and moisture.) As space-conscious Europeans know, even vegetables lend themselves to colorful geometrical arrangements featuring fruits. Hedgerow plantings of cucumbers, squash, tomatoes, broccoli, and so on can be intercropped with young berry bushes or dwarf trees. Where just one accent plant is wanted, consider a lush elderberry with ornamental foliage or a delicately beautiful Japanese plum.

There are many ways to integrate fruits into your landscape. Your own sense of proportion and design will serve to guide you, and you'll also be able to pick up valuable tips from a good book on landscaping. If you can afford it, get help from a landscape architect.

2

Planting and Caring for Fruits

The preferred seasons for planting fruits are determined by the winter hardiness of the plants or trees and the temperature extremes of the planting locale. Basically, the goal in planting is to allow as much time as possible for the young trees or bushes to establish themselves before the most stressful conditions of the year occur. In northern areas where extended periods of temperatures at $-10°F$ pose the greatest danger to trees and flower buds, early spring is generally the best time for planting deciduous fruit trees — especially for dwarfs and for the tender stone fruits. Spring planting also is recommended in colder climes for raspberries, blueberries, and for blackberries and their relatives. For such fruits early spring planting makes severe winter injury less likely by offering the advantage of a long, mild growing season that allows them to mature and harden-off before experiencing extreme temperatures. Strawberries also benefit from early spring planting in all but the mildest climates, for they thrive in moderate to cool weather, which stimulates runner production. In the northern United States, spring-planted fruits also profit from the higher level of soil moisture present then.

In warmer areas where the long, hot summers are the most stressful time for plants, most fruit trees do better if planted in fall or early winter. Fall planting allows the young stock to regain the strength lost in transplanting before its energies are called upon to produce new spring growth. The soil at that time of year is warm enough to allow good root growth before strong topgrowth gets under way in the spring.

The soil in the fall is often more workable. Fall planting avoids having to compete with heavy spring rains that often make the soil difficult to manage. Many fruit trees also benefit from the winter's alternate freezing and thawing, which helps to settle their roots. Furthermore, the foothold gained over the winter affords almost a year's head start toward bearing and allows fruit trees to withstand midsummer droughts more successfully. Plants that have settled in over the winter also start growing earlier the following spring than do just-planted fruits.

Preparing the Plant

Ideally, fruits should be planted as soon as possible after they arrive by mail or are brought home from the nursery. If the plant is bare-rooted and planting must be delayed, dig a shallow, foot-deep trench for trees (6 inches deep for small fruits) in a shady location and lay the exposed roots in it so the top of the plant emerges at a sharp angle. Then cover the roots with soil, which should be kept moist. This interim treatment, called "heeling-in," can be used for up to several weeks if necessary.

Another approach is to keep the plant in a humid cellar or shed, soaking its bare roots, then covering them with moist soil,

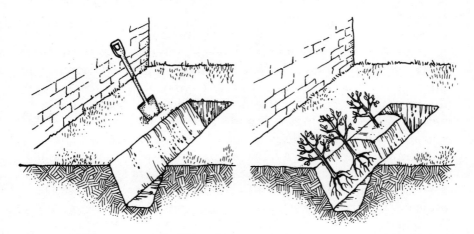

If the planting of bare-rooted nursery stock is delayed, the new plants should be "heeled-in" immediately upon arrival. Dig a temporary trench deep enough to hold the roots, in an area protected from direct sun and wind. Place the young stock at an angle in the trench (preferably with their tops pointing to the south to prevent sunburning of the trunks) and cover with soil. Water thoroughly and keep moist until they are transferred to a permanent site.

straw, or burlap. If planting is to be postponed for only a day or two, soak the plant's roots and keep it in a cool, shady spot. If the plant has a rootball wrapped in burlap, simply keep the covered roots moist until planting time.

Whether you do or do not plant right away, it is essential to keep bare roots from drying out. It is best to unpack and unwrap such plants immediately and then soak the roots from several hours to overnight before planting. You can use a slurry of 1 cup of compost per gallon of water. If you are planting a group of new fruit plants at one time, keep the roots of the vines, trees, or shrubs in plastic or under the wet burlap to prevent them from being dried out by exposure to the air, wind, or direct sun before you can set them in the ground. Dipping the roots in a slurry of clay and manure in water right at planting time will help to seal in the moisture.

Preparing the Site

Either before or after you attend to root moisture, you must prepare a suitable site for your fruit planting. Planning in advance can be helpful, especially if the soil at the intended location is not of high quality or has been used to grow corn or other demanding crops. If a sizable spring or fall planting of fruits there is antici- pated 1 or 1½ years ahead, you can plant and till under a green manure cover crop of winter rye, oats, or clover or other legumes to add nitrogen to the soil. Another option is to dig the hole for spring planting the preceding fall and make compost in it. If this is done the hole should be at least twice as wide and deep as the root spread of the plant that will go in it. If you plan to plant in the spring in an area formerly covered by sod, till the area in the autumn, digging in lots of fresh or dried manure (this is especially important if you will be putting in heavy feeders such as berries, grapes, or melons). You can also add lime at this time if a soil test indicates acidity and you do not intend to grow blueberries or other fruits favoring a soil with a low pH. In any case, work this area well before planting fruits the next spring. For fall planting where the soil is good enough to grow grass or vegetables, you can simply dig the hole a month ahead.

Whether you plant in the spring or the fall, it is imperative that the hole be deep and wide enough so that roots can be fully extended without crowding or bending. Indeed, for best results the hole should be half again as wide and deep as the roots reach when extended to their full length. A good rule of thumb for a bare-rooted tree would be a hole 3 feet deep by 3 feet wide; if there is a rootball instead of bare roots, the hole should be at least 6 inches wider in all directions. In digging the hole, be sure to shovel the topsoil and the subsoil on different heaps, so you can later place the richer topsoil, either alone or amended, at the bottom of the hole, where it can do the best job of feeding the roots.

Before You Plant

Before the tree or plant is actually set in the hole, matters of drainage must be attended to. If the substratum is clay, it may be necessary to use drainage tiles or, before digging the hole, to raise the planting area by about 2 feet using a mound of topsoil and humus with the sides sloped. If the drainage is fairly good, however, it is sufficient to put down 2 inches of gravel or a layer of rocks in the bottom of the hole or container.

The next step is to put down a layer of topsoil. Often it is desirable to improve the drainage and nutrient content of that soil by mixing it with amendments such as sand, fibrous peat, leaf mold, and compost. Shape the soil into a mound or pyramid, so the roots can be gently arranged downward over the sides of the mound and into the bottom of the hole.

Pruning at Planting Time

Before placing a bare-rooted fruit tree or bush in the planting hole, use a sharp knife or pruning shears to cut back any broken tips by ½ inch so that healthy, vital tissue is next to the soil and more delicate roots are stimulated to grow. Since about half of the plant's root structure was left behind when it was dug up at the nursery, it will also be necessary to remove an equivalent amount of top growth just before or after planting. Otherwise, the disproportionately greater amount of evaporative surface above ground will slow or stop growth, or even cause branches to die.

If the tree you are planting is simply a whip with no branches, cut off roughly the top third. If you are working with branched stock, a good rule of thumb is to remove any branches that emerge from the main trunk (or leader) at an angle of less than 45 degrees. At this time in the tree's life, you want to encourage the development of a strong system of scaffold branches to support the fruit later on. Branches growing more or less vertically will compete with the central leader, and also form narrow, weak crotches that tend to split under a heavy fruit load. Choose three to five branches that will become scaffolds and remove, completely, all others except the main stem or leader. The scaffold branches that remain should be radially distributed as evenly as possible around the tree. Head back the leader to a total length of 3 feet or so, except in the case of peaches, nectarines, and apricots, and spreading, low-growing plums, which do better with an open center. For open-center trees, cut out the entire leader.

It is not a good idea to merely tip-prune all the branches because this encourages the growth of lateral branches, which is undesirable at this stage of the tree's growth.

Pruning should be done down to buds on the outside of branches so that new growth will be outward rather than toward the center of the tree. Bush fruits should be trimmed of dead or damaged wood and short twigs, and any fruit buds removed so growth of roots and branches is encouraged. All pruning cuts should be made close to the trunk or limb, with no stubs left (these can be a ready site for disease entry).

If you plant a bare-rooted bush, take the same care that you would with any other plant to ensure that the top and the roots are in balance with each other. Remove any damaged roots and cut the top back about half. In order to encourage the development of a bush plant, pinch any blooms that appear the first year, and for the next few years the only pruning necessary is the removal of damaged or dead branches.

Positioning the Plant

After topsoil has been placed in the bottom of the planting hole in the shape of a pyramid, the bush or tree should be lowered on top of it in as straight a position as possible. If the plant is bare-rooted, gently extend the roots outward and downward over the hill of soil. Cut off any broken or dead roots. Any large root that is inextricably wrapped around another should be pruned back, so that in growing it does not choke off the other root's supply of moisture and nutrients.

If you are planting a tree with a burlapped rootball, first cut the burlap in several places—it takes longer to decompose than you might imagine, and can confine roots to their detriment.

The depth at which fruits are properly positioned varies. Bear in mind that fruits should not be planted too deep, for it's important that feeder roots grow near the soil surface to anchor the bush or tree firmly. Bush fruits or ungrafted trees are usually put in at the same depth they were at the nursery or 1 or 2 inches deeper. (To ascertain the original planting level, look for differences of color on the bark or stem.) Where grafted trees are concerned, standard sizes are always planted with the union 2 inches or so below ground for protection from severe weather and for anchorage. Dwarfs, though, should be placed so that the graft site is an inch or two above the ground so the scion cannot send forth roots and turn the diminutive tree back into a large specimen.

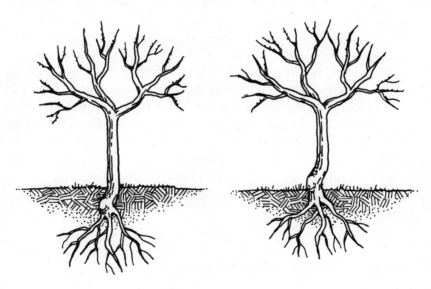

When planting a dwarf tree (right), *do not* bury the graft-union where the dwarfing rootstock joins the scion. The graft-union is easily identifiable as a discolored bump on the trunk. If the union is underground the scion will root and you will lose the effect of the dwarfing stock. When planting a standard-sized tree (left), the graft-union should be planted below ground level. Otherwise the rootstock will send up shoots that will compete with and eventually crowd out the main tree.

Settling In the Plant

When the fruit plant is positioned, an assistant will be needed to move it gently up and down and sideways while you carefully spade in more of the improved soil around the roots. This movement will help soil particles to sift snugly around all the roots. As you spade in the soil you can help this process by pushing some of it into root hollows to make sure all the cracks and crevices are filled. When several inches of soil are covering and protecting the roots, begin to stamp down on it now and then between additions of soil while your partner holds the tree in place. Tread carefully, especially if the plant has shallow, fragile roots. And do not press down too hard on soil above burlapped roots, although you should pack soil firmly around the rootball. Stamping down firmly is also not desirable if the soil is moist enough to cake. (If the soil is that wet, you would actually do best to delay planting for a few days until the soil is more workable.)

When the hole is two-thirds filled, follow some treading with a pail of water. After it has soaked in, fill the hole to ground level with the remaining subsoil. Then stamp the soil one last time, which will leave a bowl-shaped hollow a few inches deep around the plant. This depression will help the young plant or tree trap moisture over the summer. If a harsh winter lies ahead, very late in the fall mound the surface soil 6 to 12 inches up the trunk or stem and extend a slight mound to beyond the perimeter of the roots. With grafted trees, remember to remove any soil covering the graft-union before growth starts in spring, or scion rooting may occur.

At this time, make sure you have removed the wire tag put on the plant at the nursery so it does not cut into the plant as it grows.

Caring for Newly Planted Fruits

The basic planting completed, plants being staked should be fastened to the support(s) at three places if they are trees and at one place if they are bushes. The tie should not abrade the wood of the new plant and create a weak place that could snap in high wind. A handy choice is an old nylon stocking or strip of cloth which can be looped three times around the stake and once around the stem or trunk. Or the plant may be secured with sturdy wire that is passed through pieces of old garden hose.

Newly planted trees and bushes must compete with surrounding weeds and even grass or sod for water and nutrients, and

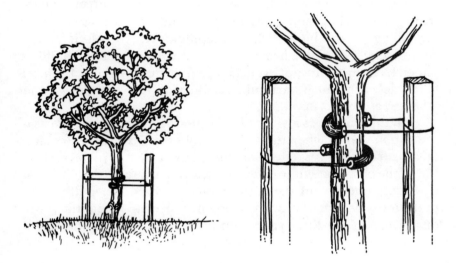

Double-staking offers good support and is especially important for dwarf trees due to their shallow root systems. Fasten the tree to stakes with wire that has been covered with pieces of rubber hose.

eliminating the competition will help the young trees along. To keep down weeds and grass, mulch or keep cultivated an area extending outward to the dripline or 3 feet in all directions from the trunk of the young tree. Another way to keep down weeds in areas where several trees are planted is to mow the orchard several times during the growing season.

If you opt for mulching, the mulch should be 6 inches deep to be effective and can consist of a layer of compost, straw, or well-rotted manure. Acid-loving plants like blueberries should be mulched with acidic materials such as pine needles, peat moss, or oak or pine bark. Mulch of some kind is also important for espaliered trees, and to protect shallow-rooted plants such as blueberries, melons, and quinces from heat and dryness. A small circle immediately around the stem or trunk should be left mulch-free since

mulch can encourage crown rot in some perennials like strawberries and it also encourages mice to take up residence. In the fall,
pull the mulch away from the trunk and install mouse guards.

The stem or trunk can be protected by encircling it with a
2-foot-high piece of ¼-inch hardware cloth that extends 2 inches
below the soil surface. If the wire circle is too much larger than the
trunk, mice will climb right inside it and spend a cozy winter
munching on the bark. Other acceptable tree guards include a
commercially available heavy plastic sleeve, light-colored building
paper, tree wrap, tape, or even a magazine, but avoid black materials which can cause harmful heat buildup during cold weather
when trees are dormant.

Young fruit trees also need to be guarded against the effects of
the sun in winter when they have no foliage. A coat of white paint
on the trunk makes effective, inexpensive sunburn protection.
Using the cheapest white exterior latex you can find (do not use an
oil-base paint), paint the entire trunk up to and including a bit
of the lowest scaffold branches. You may need to repeat the proce-
dure every two or three years, but it's well worth the time and effort involved.

Bringing Young Trees into Bearing

To ensure optimum cropping and fruit quality when the tree is
mature, it is important not to let the tree bear fruit too heavily early
in its life. Remove the fruit produced during the first two growing
seasons before it ripens. The third year the tree can be allowed to
bear a light to moderate crop. Always thin heavy crops so that
limbs do not break under the weight or distort the shape of the tree.
With some trees, dwarf apples in particular, you must pay special
attention to the leader; if it is bent by a heavy load of fruit it may
turn into a lateral fruiting branch, and the tree will be left without
its central growing branch.

Trees will usually come into bearing with no assistance needed
from you. The start of bearing is sometimes delayed when a tree
grows vigorously. In such cases, slowing the growth will encourage
fruit production, so do not fertilize the tree. Some trees may need
special treatment to encourage fruiting.

Getting Reluctant Trees to Bear

When a tree is of bearing age and does not produce fruit, there are some steps you can take to encourage fruiting. The first step would be to get a soil and leaf test done by your local agricultural extension office. If the analysis turns up no nutrient deficiencies, other measures can be taken. One is to try to spread the branches into a more horizontal position—this slows growth and promotes the development of buds. You can purchase branch-spreaders from a nursery supplier, or tie the branches at a 45-degree angle with sturdy rope or wire covered with a piece of rubber hose to prevent damage to the tree. Keep the branches in this position until a crop is set; it may take more than one season because the buds are initiated in the summer to bloom and bear fruit the following year. Branches can be spread and tied at any time, although the ideal time is immediately following the early spring pruning.

If spreading the branches fails to produce a crop, there is one further treatment that can be used on trees that are at least five

Encourage your reluctant-bearing fruit trees to set fruit by tying down their branches into a less vertical position. Increasing the angle of the scaffold branches to 45 degrees reduces vegetative growth and promotes the formation of fruit buds.

years old. This is scoring the bark, and it should be attempted only as a last resort because it can easily injure the tree. Three to five weeks after the tree has bloomed, cut through both the outer and inner bark with a sharp knife, inscribing a circle on the trunk just below the lowest branch. The only time this method should be attempted is when you would otherwise get rid of the tree.

Pollination and Fruit Set

There are some measures you can take to encourage better pollination generally in your orchard or berry patch. Growing more than one variety of a given fruit is good insurance that your flowers will be pollinated. As a temporary measure each year, you can suspend bouquets of pollinator flowers in cans of water up among the branches, and hope that the bees will spread the pollen. This approach can be very successful.

If your property is small, and there is room for only a few trees and bushes, you might want to consider top-working your trees for better pollination. This method entails grafting a branch of a pollen source onto a tree that needs the pollen, creating an all-in-one effect. This is reported to be one of the big advantages of the "five-in-one" apple trees offered by some nurseries.

After pollination, each developing fruit follows an inherited sequence of events that will eventually give it the characteristic size, shape, color, and flavor of that variety. The success of each one of these events depends on a number of outside factors — temperature, rainfall, amount of sunlight, activity of insects, diseases, and nutrition of the plants are five of the most important.

On apples, cherries, and other temperate plants, the flower buds form the summer before they bloom and bear fruit. If the fruit crop during a particular year is a heavy one, you can expect fewer flower buds to form for the following season, and the next year's crop will usually be smaller. Conversely, a light crop will be followed by more buds and a larger crop the next year. The alternate bearing that is evident in many apple trees is a direct expression of this effect of crop size on subsequent bud set.

After the flower buds have formed, make sure that your fruit trees and bushes are adequately fed and watered. This is the time of primary growth for the new buds, and if they have been well nourished they will grow strong and have a better chance of surviving the winter to develop into healthy fruit the next season. It's important not to overfeed the trees, however, and to stop all fertilization by midsummer so that growth stops in time for new buds and shoots to harden for the winter.

Low temperatures and moisture loss during the winter can damage or kill the flower buds, but the biggest threat at that time of the year is a winter thaw. Orchardists dread it because there is a chance that the flowers will open during the warm spell only to be frozen when the cold weather returns. The same sort of cold damage will be caused by a cold spell during the normal blooming season.

Once the flower bud has bloomed and been pollinated in the spring, environmental conditions again greatly affect the fertilization and fruit-setting process. Cool temperatures and bright sunshine are the optimal conditions for fertilization.

If the flowers have withstood the obstacles set in the way of fertilization, you should eventually have a quantity of plump fruit ripening on your trees and bushes. However, if there are too many fruits on one plant, there may be a spontaneous drop or self-thinning four to six weeks after bloom to lessen the competition for hormones, nutrients, and moisture. The largest, most luscious fruits come from plants with a low percentage of fruit set, so in this case, quantity directly affects quality. Keep in mind that even if only a small percentage of the fruits mature, you'll still have a good crop. A 5 percent rate of fruit set is considered adequate for apples; for peaches the rate is even less. In fact, if your trees don't drop enough immature fruit, you should thin the crop to improve its quality.

3
Pruning and Training

There are four main reasons for pruning fruit-producing plants: (1) to direct and control a plant's growth, (2) to encourage the production of flowers and, ultimately, fruits, (3) to maintain a plant's health, and (4) to rejuvenate a neglected plant. The old adage that an ounce of prevention is worth a pound of cure surely applies to pruning. In general, careful pruning during the first few years of a plant's life can encourage it to grow to a pleasing shape and stay healthy and strong and capable of producing a quality crop for many years. But always take into consideration the needs and habits of each particular tree: dwarfs should not be pruned much when young, and genetic dwarfs should not be pruned at all except to remove dead wood.

Except in the case of heading back, the pruning cuts you make on a plant or tree should stop the plant from growing in one direction and encourage it to grow in another, meaning that you can control both the size and shape of the plant. This control can be as extreme as espaliering, training a tree to grow flat along a wall, fence, or trellis, or it can be as simple as pruning to keep a tree at a reasonable height and make it easier to pick the fruit.

The ongoing process of controlling and directing a plant's growth usually begins when it is planted. When a fruit tree is planted, both the roots and top may be pruned so that they are in balance with each other. As the tree grows and matures, pruning ensures that the limbs are evenly distributed around the trunk and that they branch out at wide angles. When the tree begins to bear, further pruning creates ideal conditions for fruiting. Fruit production is increased by thinning out branches so that more sunlight and air can get to the center of the tree, promoting the development of fruit buds, and as fruits begin to develop early in the season they can be thinned so that the remaining ones grow larger.

As a plant grows, pruning to maintain its health involves removing dead, diseased, or injured stems or branches as well as unwanted growth. Dead wood in trees not only looks bad, but it also can harbor diseases and insects that can ultimately destroy a tree. Branches cracked or broken by snow accumulation and wind storms should also be pruned away. Various forms of unwanted growth can also injure trees. Branches growing into the center of the tree can rub against others and injure the bark, and since any type of damage to a plant or tree is an open invitation for disease or insects to enter, the growth of skewed branches should be kept to a minimum. Vertical-growing suckers and water sprouts, other types of unwanted growth on trees and bushes, rarely develop into healthy, fruit-producing branches and should be removed.

Each fruit has different growth habits and requires different training and pruning techniques, and the general information that follows will serve as a primer on basic pruning practices. (See Chapter 5 for details on pruning specific fruits.) Remember, too, that practice makes perfect and that you can learn a lot about pruning just by keeping your eyes open. Watch your plants grow, and observe what happens to them after pruning. You'll be amazed at how much they can teach you over the years.

What to Prune

Water sprouts grow vertically from a branch or trunk, often after heavy pruning. They seldom produce flowers or fruit and should be removed unless one is needed to replace a branch that has been injured or removed. Suckers sprout from the roots of a tree, or the base of a trunk, below the graft-union, and if they are not removed, they can eventually crowd out the main tree. Since the suckers come from the rootstock, they defeat the purpose of the grafting if they are not removed.

Buds are probably the most important part of a plant when it comes to pruning because a gardener can direct and control the growth of a plant by removing selected buds. Hormones control the growth of buds, but not all buds grow. If a bud is removed some of these hormones are lost, and this loss is a growth signal to buds lower on a branch. For example, every branch has a bud at its end called the terminal bud, from which the branch grows in length. Remove a terminal bud, and the closest lateral bud begins to grow in an effort to compensate for this loss. Lateral buds are arranged in clusters, pairs, or alternately along a branch depending on the species. Latent buds are usually invisible and grow only when other buds above them are removed.

It's easy to see how to shape a plant by removing certain buds. If you want to make a plant bushy, remove the terminal buds and new branches will grow on the sides of the original branches. If a

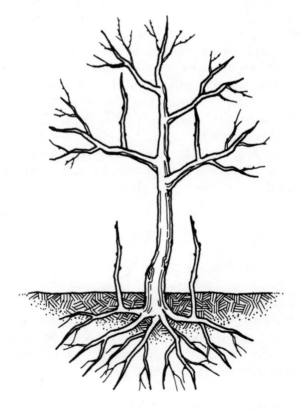

Remove all suckers and water sprouts soon after they appear on your fruit tree. Cut off the suckers from roots at ground level, and cut back all other suckers and water sprouts to the trunk, stem, or branch with a clean cut leaving no stubs to rot.

lateral bud comes out of the right side of a branch, cutting the branch off just above this bud will direct a new branch to grow out to the right. If you want an open-centered plant, prune so that all the topmost buds point outward.

How and Where to Make Pruning Cuts

The pruning requirements of plants and trees vary, but you won't go wrong if you always cut just above a healthy bud or back to a main branch, and leave a clean wound so that healing is rapid. Never leave a stub by cutting far away from a growing point. The stub will rot and invite diseases and insects. Remember always to make neat, clean cuts so the cambium tissue is damaged as little as possible and the wound will heal rapidly. Also, be careful when making the cuts that no outer bark is peeled off to expose the inner bark.

When you cut back to a lateral bud, select one that's growing outward and cut about ¼ inch above the bud in the direction in which it grows. When possible, cut from the bottom up. Smooth any ragged edges with a pruning knife. If you use a pruning saw to remove limbs, first make a shallow cut on the underside of the branch. This prevents the limb from stripping away the bark on the larger branch to which it is attached if the limb falls before it is completely cut.

The ultimate goal in removing a large branch is the same as for removing a smaller branch: make a neat cut and don't leave a stub that will rot and encourage disease. If the branch is relatively light, you can support it with one hand while you operate a pruning saw with the other. If the branch is heavy, it will have to be removed in sections, and you might also have to use ropes and slings to support and lower sections of the branch to the ground.

The first step in removing a large branch is to cut off any side branches, and then saw off the branch to within about 2 feet of the parent branch or trunk. Remove the branch in sections if it's heavy. The final three cuts are the crucial ones. First, make a shallow one on the underside of the branch about 6 inches from the trunk. This will prevent any tearing or ripping of the bark when the next cut is made. Make the second cut about an inch from the undercut, and farther out on the limb, sawing from the top of the branch. Finally, make the third cut from the top to the bottom and close to the trunk to prevent leaving a stub.

After making a cut on a large (1 inch or more) limb, the cut end should be coated with a material to prevent disease entry and encourage healing. Wound dressings or pruning paints come in aerosols and bulk cans.

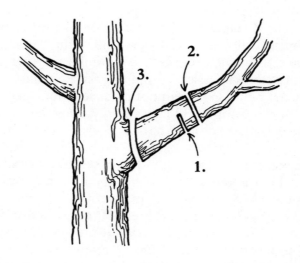

The final three cuts in removing a large branch are important to avoid ripping and tearing the bark. (1) Make a shallow undercut about 6 inches out from the trunk. (2) About an inch beyond the first cut, sawing from the top of the branch, saw through the limb. (3) Then remove the remaining stub, cutting from top to bottom, with a smooth cut.

Pruning Techniques

There are a variety of pruning techniques: pinching, top cutting, thinning out, root pruning, and rejuvenation pruning. During the summer when trees are in active growth, you can pinch off any new green growth that will break easily between your thumb and index finger. Pinching during the growing season removes unwanted growth or it can redirect growth. Pinching the tip of a main branch stimulates growth on lateral branches, and pinching the tips of lateral branches encourages the laterals to branch and the main branch to grow in length.

Top cutting, or heading back, means removing a branch back to a bud or back to a lateral branch and is done to stimulate new growth so that a plant becomes thicker and bushier. If you want a plant to grow outward, cut back to outward-facing buds so that the new shoots branch out. If you want a bushy plant, cut back to inward-facing buds. Keep in mind that cutting back is not the same as shearing, which is clipping the ends of branches without regard to cutting back to buds.

When a tree has been neglected, thinning is often the only way to invigorate it. Thinning is the removal of an entire branch back to the ground, the trunk, or the main branch. Fruit trees are often thinned so that more sunlight and air can reach the interior of the tree. New shoots and leaves are then freer to develop and fruit ripens quicker. Thinning is preventive medicine when branches that are weak and in danger of breaking are removed. Thinning in a time of drought can reduce the water needs of a plant: if there is less foliage, there is less need for water.

Fruits as well as branches benefit from thinnings. Thinning not only produces bigger fruit of better quality, it also encourages a tree to bear fruit each year. Early in the summer when the green fruit is the size of marbles, pinch off excess fruit so that there are a few inches of space between the remaining ones. This is usually done just after some of the small fruits have begun to drop naturally. Allow about 6 to 8 inches between apples, 3 to 4 inches between apricots, 4 to 5 inches between nectarines or peaches, 6 to 8 inches between pears, and 3 to 4 inches between large plums. Berries and other small fruits usually don't need thinning.

Rejuvenation Pruning

Throughout their lifetimes, fruit trees require care and nourishment; if neglected they can turn into a tangled mass of unproductive growth or become stunted and unfruitful. It is possible to rejuvenate a neglected tree by pruning, but the first step is to determine if it really can be saved. No amount of pruning will restore trees with rotten trunks or those that are extensively diseased or pest-ridden.

But if a tree appears to be sound, and if there is some potential for fruit production, then pruning might save it.

The first step in rejuvenating an overgrown tree is to remove all of the competition for water and nutrients by removing the brush and weeds that surround it. Provide adequate nourishment by topdressing with well-rotted manure and compost, and perhaps mulching. The purpose of the first year's pruning is to remove all dead, diseased, and broken branches. It's best to burn these branches since they are likely to house insects and disease. Don't remove too many limbs in the first pruning because this would upset the balance between the roots and the top of the tree. This initial pruning will stimulate new growth, and you will be able to see the following year which branches are the most vigorous and worthy of being saved. Summer pruning tends to limit growth while pruning in winter stimulates growth, so late summer might be a good time to remove unwanted lateral branches. A neglected tree might have a lot of vertical limbs (which tend to be unfruitful), so you may want to train some of them to grow more horizontally by using branch spreaders or tying them down as described in Chapter 2.

In the second year, light thinning at the top will allow more sunlight into the interior of the tree. Take care not to remove too much, or the remaining branches and fruit may sunburn on hot days. Any weak and unproductive older branches can also be removed at this time. In the third and fourth years, remove any undesirable larger branches and continue to gradually thin out other unwanted limbs. In subsequent years, if the tree is in good condition and bearing fruit, prune it as you would any other tree.

Pruning can stimulate stunted trees to grow, too. Again, the first step is to eliminate weeds and brush that compete for water and nutrients, and provide adequate nourishment in the form of mulches and fertilizers. Since a stunted tree has grown slowly there should be fewer branches to remove, but the cutting back of any new growth should stimulate the tree to grow. Thinning spurs and removing all or most of the fruit for a few years can relieve the tree of the burden of reproduction while it is growing. But thinning spurs is a delicate operation—removing too many could set back the tree too much. When the tree begins to grow and produce, follow regular pruning procedures.

Shaping Fruit Trees

Commercial fruit growers have developed three training methods for fruit trees that home gardeners can successfully adopt for their trees: the central leader, the modified central leader, and the open-center. In the central leader method (the best shape for most apple varieties), the central branch, which appears to be an extension of the trunk, dominates, and the tree takes on the shape of a pyramid. The advantage of this shape is that it creates a very strong tree, but the disadvantage is that the center is shaded and fruit production suffers if the scaffolds are not spread out sufficiently.

The three forms in which fruit trees are usually trained are from left to right: central leader, open-center, and modified leader.

To establish a tree with a central leader, prune back a whip to about 3 feet at planting time. This stimulates the buds at the top of the whip to begin growing. In the second year, select one vertical branch to be the central leader and about four other branches to form the scaffold branches. Select scaffolds at slightly different heights on the trunk and be sure they branch out at wide angles and are evenly spaced around the trunk. Wooden or metal branch spreaders may be used to ensure a wide crotch angle. Remove other branches that are likely to compete with these. In the following years, prune to direct the tree to grow into a pyramid shape, but prune sparingly until it begins to bear. Eventually you will have to cut back the central leader because it will become top-heavy with fruit and tip over.

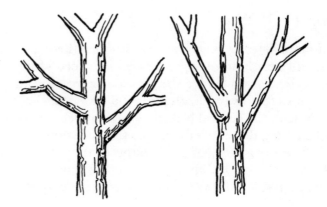

A strong scaffold system should have branches with wide angles (greater than 45 degrees) as shown on the left. Primary scaffold branches with narrow V-shaped angles (right) are undesirable as they will break under heavy fruit loads.

The open-center method is used for peaches, Japanese plums and some apricots, and other trees that don't normally reach great heights. Three to five main branches form the framework of this type of tree, which has no central leader. The branches emerge from the trunk at wide angles to allow plenty of sunshine and air into the center of the tree, which encourages the development of fruit. Care should be taken when training a tree in this method to avoid having scaffold branches that emerge from the trunk at the same height. This greatly weakens a tree and creates the possibility of heavy damage from breakage. Also, water pockets can develop between the branches and make rot more likely.

To train a one-year-old tree in the open-center method, first cut it back to 2 to 3 feet above the ground. This cut stimulates the topmost buds to start forming lateral branches. The branches should be 6 to 8 inches apart, and come out from the trunk at wide angles. They should also be evenly spaced around the tree. Remove branches that do not meet these criteria. In the second year, remove all but the three or five main branches.

The modified central leader is a compromise between the central leader and open-center methods. It's widely used for fruit trees and puts equal importance on the central leader and the scaffold branches. The central leader is allowed to grow to a height of about 5 or 6 feet, and then it is cut back. Four to six well-spaced scaffold branches are distributed evenly along the modified leader.

With all three of these methods, after the shape of the tree is established and it begins to bear, pruning to maintain the health of the tree and encourage fruit production, as described earlier, is necessary.

During the first three years, most fruit trees are pruned in essentially the same ways. But after a tree begins to bear, pruning requirements vary according to the species. (See Chapter 5 for details on best pruning methods for specific fruits.)

How to Prune Bush Fruits

Bush fruits such as blueberries, cranberries, and gooseberries are deciduous shrubs, and their growth patterns and pruning requirements are similar to those of fruit trees. Keep in mind, though, that bush fruits are trained to send up several stems, or canes, instead of developing a single trunk. Bush fruits, in general, require less pruning than other fruits, although pruning after a bush begins to bear encourages the production of quality fruit.

When the plant begins to bear, remove weak laterals, and cut back old canes to the ground to force new canes to grow. Never prune an entire bush back to the ground because the shock will be too great and the plant will die. Knowing something about the fruiting habits of each type of bush fruit will help you determine which branches to cut back.

Pruning Bramble Fruits

Bramble fruits, such as blackberries and raspberries, must be pruned regularly or the plants will deteriorate. Almost all bramble fruits have roots that are perennial and woody stems, or canes, that are biennial. The canes develop into their full height in one year, and the following year they bloom, produce berries, and die or become barren. While some canes are bearing, others are growing in preparation for bearing the following year.

The primary pruning requirement for brambles is to cut back the canes to the ground promptly after fruiting, and then to train the new canes that grow up from the base of the plant. Frequently, brambles are trained on wire trellises or poles, and depending on the variety, they can be trained in numerous ways. Canes that have produced fruit will become dry and brittle, and unless they are pruned down to the ground, insects and diseases are likely to set in. Burn these canes or haul them to a landfill.

Most bramble fruits are trained and pruned similarly. Cut back newly planted bushes to stubs a few inches from the ground. At the end of the first growing season, cut back to the ground all but the five or six most vigorous canes from each plant. Brambles have large root systems, so remove any suckers that appear. Remember, there will be no fruit produced this first year, so in order to encourage fruit production in the following year, cut back the tips of the canes to encourage fruit-bearing laterals to grow. The laterals that form can then be cut back also. As fruit develops on the previous year's growth, new shoots appear at the base of the plant. Usually about five of the most vigorous new growths are encouraged to grow and all of the others are cut back to the ground. In the fall, after the canes have fruited, cut them back to the ground, and if you are training your brambles on a trellis, tie up the new canes in preparation for bearing the following year.

The biennial canes of bramble fruits grow vegetatively the first season and then flower, fruit, and die the second year. On this principle, cut off to the ground all the second-year canes after they have fruited. Keep five to six healthy canes from the current year, cutting their tips off to encourage abundant lateral growth which will produce fruit the following year.

Decorative Pruning

Espalier is a way of training and directing the growth of a plant so that its stem and branches grow flat along a wall, fence, or trellis. Apples and pears, especially dwarf varieties, are excellent candidates for espaliering, but many different fruit trees and perennial vining fruits can be trained into some form of espalier. Training determines the pattern, which can be elaborate or simple, and pruning maintains the desired shape. The advantages of this training method, in addition to its obvious ornamental qualities, are that fruits can be grown in limited space and that the fruit is easier to reach. Also, the quality and quantity of fruit increases because the branches are exposed to more air and sunlight. Bending limbs away from the vertical means that carbohydrates manufactured by the leaves are more readily available to produce more flower buds and thus more fruit.

In cooler climates, espaliers are ideal for south-facing walls where sunlight is plentiful and the heat reflected from the walls can help in the production of flowers and the ripening of fruit. In areas with very warm summers, it is best to use a free-standing trellis because reflected heat from a wall can sunburn bark and fruit or otherwise damage a plant. You can support espaliers from pipe or wooden posts strung with horizontal, vertical, or diagonal strands of 14-gauge wire. Horizontal wires should be about 12 to 16 inches apart and the lowest one should be about 16 to 20 inches from the ground. Some gardeners use plastic-covered wires so that the branches are not injured when they rub against them. Any structure you devise should be strong enough to support the weight of the fruit, and be sure that the trunk, or stem, of the plant is 8 to 10 inches from a wall so that there is adequate air circulation, and so that you will be able to paint or maintain the wall. Use twine, rawhide, plant ties, cloth strips, or any material that will not girdle the branches, to secure them to the support structure.

For gardeners intimidated by espaliers, nurseries sell plants that have already been trained. If you'll be doing the training yourself, it's a good idea to start out with a bare-rooted whip, which will be easier to train than a more established tree. An easy and popular style is a single or double cordon in which the horizontal branches grow out at right angles from a cut-back trunk.

Many varieties of apples and pears lend themselves beautifully to this style.

Apples and pears are frequently grown in oblique, single-stemmed cordons. In this form of espalier, trees are planted and trained at 45-degree angles and secured to a support structure — although they can also be trained vertically or horizontally. In general, cordon espaliers are not well suited to the growing habits of stone fruits.

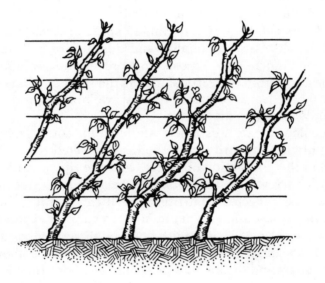

With the single-cordon method of training trees, fruit grows on dense spurs up the trunk. The trees can be trained to grow vertically, horizontally, or any angle between. This technique is useful for apples, cherries, pears, and other plants that produce fruit on spurs.

Growing plants on trellises is similar to espalier and is recommended for fruits such as blackberries and raspberries and for bush fruits such as currants. Trellises can be in the form of arbors, or they can simply be strands of wire strung horizontally between posts. The trailing canes of blackberries, for example, may be fanned out and tied to wires or they can be woven in and out among three horizontal wires of a trellis. Raspberries are usually allowed to grow vertically and then are secured to a wire

(continued on page 51)

Photo 1: Peaches are ready to pick when the last remainder of green has turned yellow. When peaches are first ripe, the flesh at the end away from the stem gives slightly to thumb pressure.

Photo 2: To make delicious spiced pears, slice a pound of fruit and add to a saucepan with ½ cup of apple or pear juice. Toss in ½ cup of sliced onions, 3 bay leaves, 3 crushed allspice berries, and a cinnamon stick. Poach the pears for 30 minutes and serve with poultry and meats.

Photo 3: Pear trees love ordinary soil because rich soil stimulates too much foliage, which can lead to a disease known as fire blight. Help discourage problems by going easy on nitrogen fertilizers. Instead, use mulch, lime, or barnyard manure.

Photos 4, 5 and 6: These apple
trees are *espaliered,* a
method of decorative pruning
that involves training trees
to grow on wires and supports.

Photo 7: Refrigerate apples and they will keep ten times longer than at room temperature. Don't pick immature apples with green bottoms, thinking they will store longer. They won't, plus they'll taste bland and starchy.

Photo 8: Of all types, red raspberries are the easiest to grow. Purple varieties are the second easiest.

Photo 9: Avoid planting strawberries where potatoes, okra, melons, eggplant, peppers, raspberries, or peaches have grown previously, because the soil may contain an organism that attacks strawberries and causes them to wilt and die.

Photo 10: This new watermelon variety, BUSH SUGAR BABY, is a bush type with short vines that take up less room in the garden than traditional varieties.

Photo 11: *Rodale's Organic Gardening* magazine recommends this unusual but effective method of supporting growing fruit trees: the screw eye in the tree is attached to a wire looped around a nearby pole. The small screw won't harm the tree.

trellis for support. In addition to supporting plants, trellises offer the same advantages of espalier in that plants can be grown in limited space and fruit production can be increased. The beauty and intricacy of both trellises and espaliers are limited only by the imagination and knowledge of the person making them.

When to Prune

After a tree or bush is planted and the initial pruning completed, regular pruning is not usually begun until a plant starts to bear. Of course, maintenance pruning to remove dead, dying, or diseased branches should occur when you spot the problem. Other pruning depends on the type of plant, when it blooms, its health, and whether it bears fruit on older wood or new growth.

Most gardeners, especially those in northern areas, do their regular pruning in late winter or early spring when plants are still dormant and there is no new growth. Gardeners living in warmer climates can prune any time in winter. Pruning in late winter is more convenient because there are no leaves on a plant, so it's easier to see what you are doing. If you don't neglect your plants, but prune them regularly, winter pruning should not result in excessive growth of wood in the spring.

In general, routine pruning in the spring involves mostly maintenance-type pruning to repair damage from storms or pests, or to remove suckers or water sprouts. It's a good time, too, to pinch off unwanted buds. Since spring is a time of vigorous growth and a period of great sap movement in trees, it is not an ideal time for heavy pruning.

Pruning in the summer usually involves maintenance pruning, but it's also a time to thin fruit for greater yields. Usually heavy pruning of fruit trees is not recommended for summer because it results in the loss of leaves whose function is to manufacture food. Sometimes pruning in late summer or fall, especially in the north, can result in winter damage because the plant does not have enough time to recover from the pruning before the arrival of cold weather.

4

Pest and Disease Control

Safeguarding fruit plants and crops is a challenge. Unlike annual vegetable crops, fruits must endure the rigors of winter, when dramatic swings of temperature or prolonged cold spells can crack and loosen bark, and weaken their defenses against infection and insect infestation.

The longevity of fruit plants also rules out or limits certain protective practices such as regular rotation of crops, planting early or late to avoid certain pests, and destroying plants after harvest to prevent continuing reproduction of pests. And soil that is occupied by a sizable root system obviously can't be deeply tilled to combat soilborne organisms. Because fruits are reproduced vegetatively they are also vulnerable to subtle long-term deterioration caused by viruses passed on in propagation material.

All this aside, however, there are ways to fight fruit pests organically. Gardeners who use these methods can expect fruit that is good to eat, although it may not be as unblemished as fruit that is sprayed with chemicals.

Preventive Measures

Good Soil Practices

For many reasons, healthy fruits are less likely to fall prey to insects and diseases than weak, malnourished crops. The quantity of nutrients available plays a critical role in a fruit plant's well-being, with unbalanced intake a direct cause of physical disorders such as little leaf, a zinc deficiency which affects various fruit trees. Malnourishment also lowers immunity by creating weakened plants that are especially attractive to certain insects. For example, many species of bark beetles and wood borers thrive on improperly fed trees while avoiding vigorous ones.

Pests such as aphids and certain diseases like fireblight are particularly likely to take hold on "unbalanced" plants whose tissues are too high in carbohydrates—a condition characterized by overly lush growth of stems and foliage. One cause of this condition is overfeeding with nitrogen. A good, healthy soil and periodic applications of compost, manure, or other organic fertilizers make such unbalanced overfeeding unlikely.

Cultural practices emphasizing large quantities of organic matter also help fight pests in some other ways. Plenty of humus in the soil works to moderate soil moisture—a fact important in pest control since too much or too little moisture can render plants vulnerable to insect damage, particularly from sucking insects which respond to changes in capillary pressure within plant parts.

Another form of cultural control is a shallow spring tilling, which helps control brown rot of stone fruits by exposing soil pathogens to weather and predators. Crop rotation of shorter-lived fruits such as strawberries and brambles to limit the buildup of insects is also helpful, as is cultivation around fruits so they do not have to compete with sod or weeds for nutrients and moisture.

Pruning and Sanitation

Proper pruning is another cultural practice that contributes to optimum plant health and makes pest or disease infestation less likely. Ideally, fruit trees should be pruned so that every leaf receives sun. In actual practice, thinning out branches will serve to open the tree, fostering the ample air circulation that will allow the tree to dry quickly after a wet period, and help foil diseases such as sooty blotch and fly speck on apples. It's been said that sunlight is the best fungicide. By opening and lowering trees by pruning, the gardener also makes it easier to reach all parts of every plant with sprays. Thinning of top growth and root pruning at transplanting time to keep roots and shoots in balance can also serve as a preventive measure against some kinds of borers. The prompt removal of raspberry and blackberry canes after they fruit is another good way to help inhibit the spread of pests.

When the wood taken off is dead or broken (conditions conducive to rots) or afflicted with canker or other bacterial complaints, pruning becomes even more urgent and a matter of sanitation. All pruned wood should be picked up and burned promptly. Where there is evidence of systemic disease—examples are fireblight on apples and pears, and mosaic disease of brambles and peaches—remove and burn the affected parts or, in extreme cases, the entire affected plant so the sickness is not carried to other fruits by insects or wind. Less drastic preventive sanitation practices include scraping off loose bark so insects can't winter over underneath it, and disinfecting garden tools and pruning wounds to limit the spread of bacteria. Wounds to the tree can be covered with asphalt pruning paint to keep out insects and disease.

It's also important to clear fallen leaves and fruit away from plants since such debris can harbor diseases like brown rot of peaches and black rot of apples, and also attracts rabbits and mice. Fruits that never developed normally should be picked from trees, as should those that begin to rot while still on the plant, for they are likely to be harboring fruit rots or insects. Healthy leaves and fruits may be composted or buried in the garden, but all pest-infected orchard trash should be burned promptly. In areas where winters are mild it is a good idea to remove not only debris but to take up mulch during the coldest month or two of the year—a practice that will help control soil insects as well as weeds.

Resistant Varieties

Circumventing pests by planting varieties bred for their ability to repel, withstand, or even damage them is a valuable preventive approach. The development of resistant varieties of fruits has proven effective against certain insects (such as the root knot nematode afflicting peach rootstocks). But resistant varieties are most important in controlling fungal, bacterial, and viral diseases. An outbreak of disease can destroy entire trees amazingly rapidly. For example, fireblight is a bacterial disease which, given the right conditions, can kill a tree in weeks. Outbreaks begin on blossoms and spread to leaves and twigs. Within days, leaves turn brown, wilt, and shrivel. Droplets of sticky ooze appear on the leaf stems.

Twigs turn a dark, purplish brown, take on a sunken appearance and droop, sometimes with an upturned tip. Infection can travel rapidly down the stem, and even into the trunk, where it kills the tree. Trees are most susceptible when they are making vigorous growth — one reason to avoid overfertilizing them. As the weather warms and growth slows, the disease may become dormant, leaving black-rimmed patches of shrunken dead bark. The next season it may revive, sometimes producing ooze from the cankers, which insects pick up, spreading the infection to new sites.

Powdery mildew, a fungus, first spreads a grayish white substance that looks like felt on the undersides of leaves. The leaves soon curl or crinkle, become brittle, and die. When it infects one-year-old twigs, it overwinters there, and will infect the tree again the next season. It seriously weakens a tree.

The one drawback to resistant varieties is that the painstakingly bred varieties eventually lose their resistance as the target insects and diseases mutate to get around it. However, this process takes many years. Along with biological control, resistant varieties may be our best hope of combating pest and disease problems in coming years.

Just a few of the fruits resistant to one or more common pests are Concord grapes, which are more resistant to black rot than some others, and the scab-resistant apples GRIMES GOLDEN, LIBERTY, PRIMA, PRISCILLA, and SIR PRIZE. Other resistant varieties are described in nursery catalogs of fruit specialists and in publications coming out of agricultural experiment stations, which sometimes provide free scion wood on request.

Biological Control

The use of living organisms to control pest animals or plants is steadily increasing in importance and holds great promise for the future. Home fruit growers are increasingly coming to rely upon natural predators or parasites to keep pests at acceptable levels. Among insect predators, those most valued at present are the ladybug, which destroys aphids, mealybugs, red spider mites, and scale insects; the larva of the *Trichogramma* wasp, which consumes moth and butterfly eggs; and the green lacewing larva, which demolishes aphids, leafhoppers, mites, thrips, and some caterpillars. All of these beneficial insects can be encouraged by organic methods of fruit culture and can be purchased by mail.

Larger predators, such as most birds, are also valuable in residence in the fruit pests. You can invite bird predators to take up residence in the fruit garden by providing winter feeding, water, materials and sites for nesting, and landscaping that appeals to them either as food or nesting material. Perhaps the single most important step you can take to lure bird predators is to create a hedgerow effect with shrubs and/or trees. A hedgerow area can boost your bird population by as much as five to ten times.

Parasites also play a major role in biological control and are considered more valuable than predators against some pests because they attack that specific pest in preference to others, developing either internally or externally on it. Often, parasites lay their eggs in the grubs of pests, later hatching larvae that consume their hosts. Insect parasites include many flies and wasps. Beneficial wasps can be attracted to your orchard area by nearby plantings of anise, carrots, coriander, dill, fennel, parsley, and parsnips that are allowed to flower.

Microbial parasites are diseases used either to kill pests or to make them more vulnerable to other controls. Although research in this mode of pest control is only now starting to gain widespread acceptance, environmental problems posed by pesticides have spurred the USDA into more research. Thus far, only a few microbial agents are on the market, notably milky spore disease *(Bacillus popilliae)*—sold under the brand name Doom—which wipes out Japanese beetle grubs, and *Bacillus thuringiensis*—sold as Dipel, Thuricide, and Biotrol—which targets moth and butterfly larvae.

Mechanical Controls

Various traps baited with insect sex attractants (pheromones) are now on the market. These can be used to monitor oriental fruit moth, codling moth, and various leafrollers in order to determine the best time to use organic sprays or forms of biological control. Other commercially available bug traps use light as the lure. Some light traps also feature an electric grid or electrified revolving wires, but these sizzle insects indiscriminately and keep the user from evaluating whether pests or harmless, perhaps beneficial insects are being caught.

There are many kinds of insect traps that you can make at home. A simple light trap may be built by hanging a light bulb with a reflector over a shallow pan of water topped with a thin film of kerosene. Another visual trap may be directed against apple maggots, which are irresistibly drawn to croquet balls painted red. These spherical traps should be coated with a sticky substance such as Tanglefoot or Stikem and hung six to a tree, where they can reduce maggot damage by up to 90 percent.

Fruitflies can also be caught in glass gallon jars containing a pint of vinegar, while a 9-to-1 mixture of water and molasses placed in the bottoms of paper cups will trap codling moths. The cups should be suspended in the trees at blossom time, for the moths emerge as the petals fall.

Where traps are inappropriate, fruit gardeners can safeguard their harvest and plants by setting up physical barriers or repellents. Tree guards made of ½-inch wire hardware cloth work well against mice, moles, and rabbits. Deer can be kept away by woven-wire fences 6 to 8 feet high. A simpler deer repellent is a bag of human hair (the scent of which frightens deer) suspended in each tree or at 20-foot intervals in perimeter trees.

Where birds are the problem, scare devices sometimes do the trick. Wind-animated scarecrows, suspended in the tops of the trees, will confound the birds for a few days; then you should switch your defense to aluminum foil strips, shiny aluminum pie pans, or artificial snakes made of old garden hose, coiled and tacked to a 24-by-24-inch piece of plywood. Place this in the treetop so it is visible from a bird's-eye view. If your neighbors are a distance away, noisemakers are good bird repellents in some areas. Looping black thread back and forth over the tree creates a barrier that, unseen but felt by the birds, frightens them away. Whatever you do, don't expect to keep all the birds out all of the time. Try several approaches simultaneously, and change methods every few days as the birds become used to each previous method.

For fruit-bearing trees and bushes that are especially attractive to birds, a more painstaking type of protection may be necessary. In such cases, the only surefire protection is to erect a barrier in the form of screen houses or cages that cover the entire tree. Or cheesecloth can be draped over bushes and fastened with clothespins several weeks before the fruit ripens. To protect trees, a better material to use is ¾-inch nylon netting which readily lets in light, water, and air. Tie the netting together below the branches. These covers are difficult to use on large trees, but they are 100 percent effective.

The stems of young trees vulnerable to borers and climbing pests can be wrapped in burlap, then painted over with whitewash. To deter slugs, scatter wood ashes around the bases of plants or young trees.

Developing fruit can be protected from birds, molds, and insects by covering it.

Organic Sprays

Sprays of the organic kind work either by repelling pests or killing them on contact. The best-known contact type of spray is dormant oil, which is applied to trunks and branches during the tree's dormant period. Dormant oil is a highly refined, light petroleum oil; apply it on a windless, frostless day (with a temperature of 40°F or above) when buds are just beginning to swell but have not yet turned green. Soil sprays can be used against aphid eggs, aphids, mealybugs, overwintering red mite eggs, pear psylla, scale, thrips, and whiteflies. Dormant oil preparations are available in an emulsified form that's mixed with water, or one can be concocted at home by mixing 2 quarts light motor oil and ½ cup of liquid detergent. One part of the oil/soap mixture should then be mixed with 20 parts of water as needed and used immediately—that is, before it separates—as undiluted oil can damage plants. Spray all parts of the tree to be really effective. The oil must coat the overwintering insects or eggs to be destroyed, because it kills by asphyxiation.

Other organic contact sprays are derived from plants. Remember that these botanical insecticides kill beneficial insects along with the pests, so they are best used only as a last resort. If you have only a few trees, it might be better to use botanical insecticides by dipping heavily infested branches directly into the solution, instead of spraying. Sprays of pyrethrum, rotenone, and ryania are relatively effective against a wide variety of soft-bodied insects such as aphids if they are applied frequently. Ryania also works against codling moths and rotenone is fatal to pear psylla. Nicotine solutions will also poison aphids. These botanical insecticides and other plant-derived sprays doubtless are effective because of the highly potent insecticidal compounds in the plants used. When a volatile oil from a plant containing the antipest substance is used directly, the results can be striking.

A sulfur spray can provide effective control of fungal diseases like brown rot on cherries and peaches. When the weather during blooming and just before harvest is warm and wet—conditions favorable for the fungus—a spray made from ½ pound of powdered sulfur to 10 gallons of water when applied to the trees will retard the growth of the fungus. Sulfur will not affect bees, beneficial insects, or people.

5

How to Grow a Dozen Favorite Fruits

Apple

There are hundreds of apple varieties available, offering a choice of color, flavor, and season of maturity. Try planting green, yellow, and red apples, some early, some midseason, some late. Grow some for eating, some for cooking or storing. Diversity reduces the risk from frost and disease and prolongs the harvest.

Most apple varieties are self-fruitful, except for RED DELICIOUS, the WINESAPS, RED GRAVENSTEIN, RHODE ISLAND GREENING, and the newer triploid crosses like MUTSU and SPIGOLD. When planting any of the STAYMAN group (STAYMAN, TURLEY, ARKANSAS BLACK TWIG, WINESAP), a third variety should be planted or the pollinator for the STAYMAN will not set fruit. The others listed all require at least two varieties for cross-fertilization. One of the best pollination choices is GOLDEN DELICIOUS, which is somewhat self-fruitful and pollinates almost every other variety. WINTER BANANA is another good pollinator, especially of RED DELICIOUS. Even self-fruitful trees will profit from being planted with a companion tree.

Any good garden soil is suitable for apple trees. They tolerate a wide range of soil types, from sandy to clay, as long as the ground is well drained. A pH of 6 to 7 is best.

Apple trees are available in standard and dwarf sizes. The most popular dwarfing rootstocks for apples are the Malling or Malling-Merton Series from the East Malling Research Station in Kent, England.

Root weakness is always a result of the dwarfing process. Some dwarf roots do not anchor a tree well, or the roots may sucker excessively. Trees on some dwarfing rootstocks (EM XXVI or EM VII) need to be staked in their early years, and EM IX trees always require stalking or trellising. Dwarf trees tend to grow more upright than standards and therefore often need more training to spread the branches.

Most backyard growers buy grafted trees to set out. Dwarf varieties begin to bear in one to three years. Standard varieties begin to bear in five to ten years.

Trees can be planted in the spring or fall, during dormancy. Growers in regions of colder winters usually plant in the spring. Growers farther south plant in the fall.

The planting hole should be large enough to accommodate the roots when they spread out. Depth of planting depends to some extent on the rootstock. Most trees on dwarf rootstocks are budded high to allow planting much deeper than they grew in the nursery. This encourages more rooting while maintaining the dwarfing characteristics.

Do not put soluble fertilizers, even of an organic nature, in the planting hole because they could burn the roots. Cover the roots with the topsoil from the site and compost, and put the subsoil dug from the hole on top. Mulch around the tree with 2 to 4 inches of organic material, but don't pile it against the tree. Water well, to settle the soil around the tree roots. Mulch in a normal year will hold in enough moisture after that initial watering to keep the tree growing between rains, but if dry weather occurs, keep watering through the mulch. Once a week, give the tree a good soaking.

Keep potash and nitrogen in balance in the apple tree's soil. If excess nitrogen is available, the tree will become too lush in its growth, will be more prone to fireblight attack, and will fail to "harden" before subfreezing temperatures begin. To make sure that potash and nitrogen are present in about the right amounts, mulch in the summer with barn manure into or over which you sprinkle 8 to 10 cups of unleached wood ashes per year. Don't overdo it, or the pH will rise. In midsummer, add to the mulch several inches of clover clippings from the lawn, or some other legume. The apple tree's phosphorus requirements are much smaller. An annual application of 3 cups of bone meal per tree, which will become available for plant food only slowly, is adequate.

An apple tree's growth habit depends on its variety, the rootstock to which it is grafted, and sometimes the peculiarities of the individual tree. Most apples are trained to the central leader shape.

You can prune apples at any time, but beginners should stick with the late dormant season, a few weeks before the buds open.

Common insect pests are various aphids, apple fruitworms, apple maggots, codling moths, European apple sawflies, European red mites and other mites, flatheaded borers, leafrollers, plum curculios, roundheaded borers, and white apple leafhoppers. Common apple diseases are bitter pit, black rot, cedar-apple rust, fireblight, fly speck, powdery mildew, scab, and sooty blotch.

Only minimal winter protection is necessary for hardy apple trees. Young tree trunks should be wrapped to discourage mice and rabbits. To prevent sunscald, paint the trunks with a cheap exterior white latex paint. Keep deep snow trampled down around trees or shoveled back.

Experience is the best teacher in learning when to pick an apple at its peak. For storage, apples are usually picked before they are fully ripe. In general, an apple that will come from the tree with a slight upward lift and twist, the thumb of the picking hand flicking loose the stem at the same time, is ripe enough. Another indication of an apple nearing ripeness is the darkening of its seeds.

Apricot

A nutritional bonanza, fresh and dried apricots contain large quantities of vitamin A, vitamin C, iron, traces of B vitamins, and minerals. Apricots ripen from early June to early August.

Give very careful consideration to the apricot's site. Though its blossoms are no less hardy than those of other stone fruits, they are the first to bloom in the spring and are very subject to frost damage. In climates prone to spring warming trends followed by freezes, plant apricots in an area that warms slowly in order to retard bloom as long as possible. A north hillside is ideal if it's not subjected to frigid, drying winds all winter, which could kill buds. It's not uncommon for even a well-tended apricot to freeze out two years out of three, but the reward in that third year is worth the wait. In addition, the tree itself is attractive and eventually grows into a fine shade tree.

Apricots tend to be quite deep-rooted (up to 16 feet) so areas with a heavy hardpan should be avoided. Russian types are somewhat more bud-hardy than peaches or nectarines. Winter chilling requirements are similar to those of peaches, so barring problems with spring frosts apricots will thrive in peach regions. See Peach for additional site selection guidelines.

Most varieties are self-fruitful, except RILAND and PERFECTION, but will not set full crops in single-tree plantings.

On its own root or grafted to a selected apricot rootstock, the tree will do best on sandy soils and have a high degree of nematode resistance. On peach rootstock, the fruit will ripen a few days earlier than on other roots, but some tree vigor will be sacrificed.

A 3- to 5-foot seedling, ½ inch in diameter a few inches above the graft, is ideal planting size; it should have no dead shoots or areas of damaged bark.

Apricots should be treated exactly like peaches at planting time except for placement of the trees. Since they will be larger trees, apricots should be spaced 30 feet apart.

Fertilize an apricot tree like a peach, but be aware that an apricot's nitrogen needs are barely half as much. Overfertilization with nitrogen causes fruit to drop while still green, as well as causing a condition known as pit burn in which the flesh around the pit turns brown and soft. A dusting of lime, wood ashes, phosphate rock or other mineral dust, and an inch or two of barnyard manure spread under the dripline each winter will supply needed nutrients.

Thin varieties that overset badly, leaving one fruit every 3 to 4 inches. This system will yield the best crop and not overtax the tree's capacity, which could reduce hardiness for the following winter.

Lack of water will cause small, mealy apricots that are best suited for the compost heap. If the weeds near the tree wilt shortly after sunrise, the soil may be reaching dry levels and will need watering two or three times weekly at a rate of 25 gallons each time until rains return. A prolonged drought can delay hardening of the wood in the fall, leaving the tree more susceptible to winter injury. Mulching heavily will help conserve moisture and hold back weeds.

Follow the pruning guidelines listed for peaches if the variety is a spreading type and to be trained to the open-center shape.

For more upright-growing varieties, use the modified central leader form. Heading the leader several feet above the main scaffolds in the third year after planting will encourage growth of secondary scaffolds. Prune lightly or fruiting will be delayed several years and the yield will be reduced.

On the main and secondary scaffolds, the fruit will be borne on short one-year spurs and on the tips of last year's growth. They need plenty of sunlight to grow and remain healthy. After three or four years these spurs and branches will stop producing fruit, making it necessary to train new shoots to develop into fruiting limbs. If the center of the tree receives inadequate sunlight, the new spurs will be weak or nonfruiting. It is important therefore, once bearing begins, to prevent the foliage from becoming too dense by thinning out old wood. Decreased fruiting in the inside of the tree indicates that pruning has not been vigorous enough to allow sufficient sunlight to penetrate. On the other hand, too-vigorous pruning severely reduces the crop. Gauge your pruning by the health of the buds in the tree's center.

A mature tree may yield 200 pounds of fruit. Major varieties tend to bear only every second or third year. Thinning and careful pruning may reduce this tendency, but there will still be noticeable differences in the size of the crop from one year to another.

Apricots suffer the same insect and disease problems as peaches. See that entry for information.

When apricots are fully ripe they will drop from the tree. If you can't wait that long at least allow them to turn from green to orange — the first sign of ripening — and let the flesh soften a little. At this firm-ripe stage, fruit will keep a week or so in the refrigerator in a moisture-tight container. Fruit can complete ripening at room temperature, but will not attain tree-ripened sweetness. For canning or drying pick a few days past firm ripe to catch the highest sugar level.

Blackberry

There are two types of blackberries: erect and trailing. Erect blackberries have canes which may reach 5 feet in height, are four-sided, and can stand without support. The trailing type has canes that can reach 15 feet in length, which trail over the ground unless supported. It is grown mainly on the West Coast and, except for the dewberry, which grows in the Southeast and as far north as Pennsylvania, it is less hardy than the erect type.

The blackberry is a bramble fruit which produces new canes by suckering freely from its roots. Without bees, which help pollination, the crop may be half or less of the normal quantity. The fruit ranges in size from ½ inch on small wild plants to 1¼ inches on the boysenberry, grown in California. The berries and the green core to which they are attached come away together when the berry is picked, and the eating quality of the berry partly depends on the core being small and soft.

Blackberries grow well in most soils, except those that are poorly drained or acid. The southern blackberry, or dewberry, will grow in dry, rocky soil not tolerated by other blackberry varieties.

Nursery plants are started from root cuttings and sold when they're a year old. They should be planted 1 to 3 feet apart in the row, in the spring. Within two years the row will fill in as the roots spread and send up new canes. Whether the plants are trained in a row or in hills, the canes which come up where they're not wanted must be pruned away underground. A mulch can increase yields by three to five times.

In the spring and early summer, blackberries send up suckers from the roots, which grow into canes and then go dormant during the winter. The next season they bear fruit and die. Blackberry pruning extends over both seasons. When the first-year cane of erect varieties reaches 3 to 5 feet in height, prune back the tip 3 to 4 inches. Topping stimulates the cane to branch, and the more branches, the more fruit will be produced next season. The next season, as soon as the cane bears fruit, prune it at ground level, then remove and destroy it to prevent disease.

Trailing blackberries are usually allowed to run along the ground the first season. The canes, which can reach 15 feet in length, should be pruned to 10 feet in late winter, and then arranged on a 1- or 2-wire trellis in loops and curves to expose as many leaves to the sun as possible. As with the erect types, canes which have borne fruit must be pruned off at the ground and destroyed immediately after harvest. Dewberries are sometimes pruned off completely after harvest—first- and second-year canes alike—because the long season in the South affords enough time for regrowth, and the plants seem less subject to disease when forced in this way to start fresh in late summer.

Red spider mites can reach damaging numbers in hot, dry

weather. A forceful spray of water under the leaves will dislodge many of them. The several kinds of borer that attack the blackberry can be kept in check by removing canes that wilt or topple over near their tips.

Blackberries are subject to several virus diseases. When symptoms appear, such as stunted new canes, or withered yellow stunted leaves on second-year canes, the diseased plant must be uprooted and destroyed, or aphids will carry the virus to the other plants. Buy stock that is certified virus-free.

Avoid planting blackberries in plots that were used to grow eggplants, peppers, potatoes, and tomatoes within the previous four years. These plants carry verticillium wilt. If the leaves on a blackberry wilt, starting from the ground up, the cause is usually verticillium, and the plant must be uprooted and destroyed.

Orange rust is the chief blackberry disease. It causes orange blisters on the undersides of leaves. Infected plants must be uprooted and destroyed. The best prevention for this and other blackberry diseases is to keep the canes from growing too thickly together.

Fresh blackberries keep only a short time under refrigeration, but they may be frozen for long-term storage. The frozen berries, however, lose much of their quality.

Blueberry

Although now a familiar fruit, the blueberry was first brought from the wild and domesticated only in this century. It is now grown widely and its cultivation is spreading rapidly into new areas, mainly because the two most important species, the highbush and the rabbiteye, have different natural ranges and soil requirements that, between them, cover a large part of the United States. Both are vigorous and productive plants, and many fine varieties are available, with new varieties appearing every year for the rabbiteye, which has been in cultivation for just 40 years.

The highbush, the usual blueberry of the supermarket, grows naturally from Florida to Maine and Michigan, wherever the soil is highly acid and well drained, and commercial growers have extended its range as far as Missouri and Arkansas. The rabbiteye is native to the Southeast and tolerates heat and dry soil better than the highbush. It also grows on less acid soil than the highbush.

Though several other species of blueberries are sold, they are all undomesticated and gathered from the wild.

Both the highbush and the rabbiteye are woody deciduous bushes with smooth grayish bark and small, glossy, dark green leaves. The highbush is the smaller, growing to 15 feet in height, while the rabbiteye may reach 30 feet. The highbush needs more winter chilling than the rabbiteye, and though more hardy, will grow only as far north as the peach will. At least two varieties should be planted together to help cross-pollination.

Highbush berries must have acid soil—preferably a pH just below 5. The rabbiteye will tolerate a higher pH than the highbush—to 5.5 or more. Both need well-drained soil, but tolerate heavy clay and sandy soil provided they contain a good deal of organic matter. Given good, acid soil, both are relatively pest and disease-free.

Both rabbiteye and highbush are planted in the spring, usually as two-year-old plants. Adding peat moss and leaves, bark, or sawdust from oaks will lower pH slowly, and in a year or two make most soil acid enough for blueberries. They also supply good drainage, which blueberries demand. But if the soil pH is 6.5 or above, it can be made acid enough only with a great deal of labor. Give the plants an acid mulch, like oak bark or leaves, and be sure to keep the soil moist. The mulch will be enough food for the plants. Highbush plants should be at least 6 feet apart, and rabbiteye plants 7 to 8 feet. It's a good idea to plant at least two varieties together, making sure they ripen at different times.

Pruning begins at planting. The new plants should be cut back by one-fourth and any low, bushy growth near the base of the plant pruned away. After two or three years of growth, regular pruning can begin. On upright bushes, one goal of pruning is to open the center of the bush. On spreading bushes, one goal is to remove the low, shaded branches. But the main goal of pruning is to keep the bush well supplied with young, fruitful branches. Blueberries bear fruit on wood that formed during the previous season, and they bear best on young branches which are sending out vigorous fruitful twigs each season. By the time a branch is six years old it has lost its vigor, grown twiggy, and has few fruit buds. Prune these old canes and see that they are replaced with new growth coming from the roots. For thinning, take twiggy branches off newer canes. If shoots appear late in the summer, prune them.

The blueberries in home plantings have few pests and diseases. If the fruitfly appears, rotenone in several applications between June and harvest will help keep the fruit free of maggots. Other pests are cherry fruitworms (¼-inch-long red worms), plum curculios, and cranberry fruitworms, which enclose a cluster of berries with their webs. If these pests reach damaging numbers, it will help to remove the mulch and keep the soil cultivated for a season. Give blueberries a dormant oil spray to control scale.

Depending on the region and the variety, the highbush ripens from as early as late May in North Carolina to as late as mid-September in Michigan. The berries change from green to red to blue as they ripen. The berries ripen over several weeks and continue to grow larger after they have turned blue. It is best to pick several times, starting six days after the berries turn blue.

The rabbiteye ripens later than the highbush wherever the two grow together. The season starts as early as May in Florida and extends as late as August in North Carolina. Often berries in the same cluster ripen at different times, and the harvest can extend over several weeks. Begin picking one or two weeks after the first blue color appears.

Cherry

Sweet cherry trees, growing to 80 feet in height, are divided into two groups according to the characteristics of their fruit: the soft-fleshed, heart-shaped Hearts or Geans (GOVERNOR WOOD, BLACK TARTARIAN) best suited for home use, and the more common Biggarreaus (BING, LAMBERT, SCHMIDT), which are sweet, crisp-fleshed, and have a storage life of a few weeks.

Sour cherry trees seldom exceed 35 feet and are more bushy and slightly more tolerant of insects and diseases than sweets. Though fewer varieties of sour cherry are available, they also fall into two groups: the red-skinned, yellow-fleshed amarelles (MONTMORENCY) and the more acid, dark-skinned morellos (EN-GLISH MORELLO) that grow as large bushes rather than as trees.

Sweet cherries produce their first crop 3 or 4 years after planting; sours, 2 or 3 years after planting. Crops increase for the next 10 to 15 years as the tree grows. A mature sweet cherry will yield 3 to 4 bushels; amarelles, 2½ to 3; and a healthy morello, 2 bushels.

Though as blossom-hardy as peaches, cherries may bloom a few days earlier and can get caught in spring frosts. To delay bloom, site them where they'll warm slowly in the spring; a north hillside is a good choice. If winter-damaged, cherries recover very slowly. Sweets are very intolerant of heavy, poorly drained soils. Otherwise, the suggestions for a good peach site will satisfy any cherry.

Sour cherries are self-fruitful and will set full crops on lone trees, but they should not be relied on to pollinate sweets since the bloom periods seldom overlap. Cherries are possibly the only fruit with more complex pollination guidelines than the plum. All varieties, except STELLA, are self-unfruitful; single trees will bloom profusely but set no fruit. The obvious solution is to plant some other variety, thus providing viable pollen for cross-pollination. This solution is complicated by the presence of incompatability groups. For example, BING/LAMBERT/NAPOLEON/EMPEROR FRANCIS all have viable pollen, but will not pollinate each other — probably due to common ancestry in breeding programs. Other incompatible groups are: WINDSOR/ABUNDANCE; BLACK TARTARIAN/BLACK EAGLE/KNIGHT'S EARLY BLACK/EARLY RIVERS; NAPOLEON/CENTENNIAL; ELTON/GOVERNOR WOOD/STARK'S GOLD; VAN/VENUS/WINDSOR; VIC/HEDELFINGER; HUDSON/GIANT/SCHMIDT/URSULA; VEGA/VISTA/SENECA; and SCHMIDT/ORELUND. For successful pollination, varieties from different groups must be planted and they must have overlapping bloom periods. Consult nursery catalogs or your local extension office for more information on suitable local varieties.

With a slightly longer chilling requirement (1,100 to 1,300 hours) than peaches, most cherries will grow best in the northern parts of peach-producing regions. The hotter, more humid summers to the south favor the brown rot fungus which can ruin an entire crop in a few days. Sweets are about as winter-hardy as the hardier plums, while sours can be grown farther north, well into apple country. One-year-old nursery stock 4 to 5 feet tall and ¾ inch thick just above the bud-union is best for planting. If two-year-old stock is available, it should have several branches indicating vigor and health, while whips of this age should be left in the nursery.

Cherry buds open very early in the spring and moisture-demanding foliage develops soon after. If the newly planted tree has not grown enough roots to meet the demand for water, the tree may die quickly. In warmer areas, fall planting is suggested, while in more northern climes where severe cold could damage the young tree, early spring planting on ground prepared the previous fall is best. Follow planting guidelines for peaches, spacing the trees 15 to 18 feet apart for morellos, 22 to 25 feet for other sours, and 35 to 40 feet for sweets. If possible, plant cherries where no peaches or cherries have grown before to avoid nematode problems. Cut the newly planted tree back to a height of 3 feet to induce branching to form scaffolds.

If treated like peaches, though fed less, cherries will thrive. Sweets are very drought-sensitive. All cherries respond well to watering, but in the eastern and midwestern United States, the water needs of this early-maturing crop are usually satisfied by winter and spring rains.

Bare, cultivated ground under the trees may help counter the effects of spring frost by soaking up heat during the day and slowly radiating it back at night. Heavy mulching with any coarse organic material is highly recommended, but wait until all danger of frost is past and keep the mulch a foot or two away from the trunk to discourage mice.

Use the length of a season's shoot growth to determine the tree's vigor: 12 to 24 inches is good for sours, 26 to 36 inches for sweets before bearing begins. When trees are beginning to bear but are still growing, 8 to 12 inches of growth is enough, dropping to 4 to 8 inches when trees are in full bearing. In humid areas, where cherries are grown in lighter soils, deficiencies of magnesium (indicated by yellow, stunted leaves) and/or potassium ("scorching" of leaf edges, older leaves first) may occur, although a manuring program should be adequate.

Those varieties of sweet cherries that taste best are also most prone to cracking. When rains occur within a week or so of harvest, the fruit may absorb too much water and burst. If heavy rains are predicted near harvest, you can reduce the amount of cracking by applying a solution of ½ cup liquid dishwashing soap to 10 gallons of water.

More than other fruit trees, sweet and sour cherries will grow into a productive shape with little pruning. Prune sweet cherries only to remove dead or broken limbs, to thin out shoots to allow sunlight to penetrate, and to eliminate cross limbs.

Sour cherries have a more spreading growth habit and are easily trained to the modified central leader shape. Avoid heading cuts and heavy pruning on sours which will stunt the tree and delay bearing. Don't attempt to thin the crop. It will not affect size.

Insects and disease can be troublesome, but the biggest challenge to harvesting a crop of red sweet cherries is birds. Not content to eat a whole cherry, they take a few pecks and can ruin the crop as fast as it ripens.

Curled leaves on shoot tips will reveal the black cherry aphid when unrolled. More of a problem on sweets, where a heavy infestation can stunt a young tree, it can be kept under control with rotenone until the predators move in. Tent caterpillars, though they will eat other fruits, prefer cherry foliage and can strip a branch or young tree in a few days. In the spring, remove any of the egg masses, found as a hardened gray foam surrounding small twigs. Other insect problems are borers, cherry maggots, and scale.

Brown rot fungus, the bane of stone fruit orchards, can be a very serious problem in cherries, especially sweets. Another fungal disease, cherry leaf spot, attacks the leaves, producing ⅛-inch purplish red spots that develop a pink cushion of infecting spores on the lower leaf surface. Both diseases can be retarded by a light sulfur spray. Sour cherries may develop a condition known as "yellows," a virus-induced speckling of the leaves. Increasing the nitrogen fertilizer to affected trees will help.

When the fruit reaches its full skin color, pick a few of the best colored by pinching and slightly twisting the cluster of stems. If they're sweet, juicy, and a bit firm, keep picking those of similar color. If you like them sweeter, try again the next day. There is no standard for picking once full color is reached and, like other stone fruits, cherries get sweeter and larger the longer they hang on the tree, until they begin to soften.

They will keep several weeks in a moisture-tight container in the refrigerator, but are best right off the tree.

Cranberry

Cranberry plants bloom in spring or early summer. Berries are bright green at first, but turn bright red by harvest in October. A good source of vitamin C, the berries are tart and rarely eaten raw.

The cranberry grows on sandy or peaty acid bogland. Having no root hairs, the plant depends on a symbiotic fungus to provide food and water from the soil, and does not tolerate dry conditions in the top 6 inches of soil. Most commercial bogs are leveled and diked on the sites of natural bogs of leather leaf, brownbush or peat, with water at hand for flooding (a practice used to control insects at certain times), to protect the plants from frost, drying winter winds and summer heat, and sometimes to help in harvesting. The water is also used for irrigation. Some West Coast growers, however, use sprinkler irrigation to give the plants the moisture they need, and grow their plants in ordinary acid soil rather than natural bogs. A heavy peat mulch protects these plants during winter.

The flowers are self-pollinating, but when honeybees and bumblebees visit the plants, there is an increase in the size of the crop and the berries.

Cuttings are planted on 18-inch centers in peaty or sandy soil (pH 5). The ground must be kept weed-free for several years, until the stems cover it and upright shoots have made a thick stand. The plants use only the top 6 inches of soil, which must be kept constantly moist. When the soil is frozen, the plants, being semievergreen, can be dried and killed by winds. To counteract drying, bogs are often flooded in winter.

The main commercial cranberry varieties are EARLY BLACK, HOWES, SEARLES JUMBO, MCFARLIN, BEN LEAR, CROWLEY, and STEVENS. There are local favorites in New Jersey, Wisconsin, and Oregon.

False blossom, a virus disease spread by leafhoppers, nearly destroyed the commercial cranberry industry in the 1920s. Three varieties of cranberry, BECKWITH, STEVENS, and WILCOX, seem less attractive to the leafhopper than other varieties and so escape false blossom more often than other varieties. The cranberry is subject to several fungal diseases like red leaf spot, but they are rarely damaging to the crop.

The cranberry fruitworm, which bores into the fruit, eats the seeds, and then exits the fruit, can cause the loss of as much as one-third of the crop in some years. The black-headed fireworm, a small, brownish leaf-feeder, causes leaves to brown as though singed. Flooding is the main natural control for these pests.

Melon

Melons that grow in the United States include the muskmelon or American cantaloupe, the European true cantaloupe, and winter melons, including honeydew and casaba. Melons can be netted or smooth-skinned, ribbed or furrowed, and green, light brown, pale orange, or yellow, with inside flesh ranging from white to green and through every shade of orange and yellow. Some melons have a pronounced musky scent, while others are scentless. Some are sweet, some bitter in taste, and many insipid.

Melons bear for one season only and must be replanted annually in temperate areas. The fruit ripens from mid to late summer or early autumn, depending on variety.

In northern temperate areas melons are not an easy crop. They are susceptible to many diseases and afflictions, including rots, mildew, and mosaics. Their seedlings are easy prey for aphids, damping-off fungus, and squash bugs, and insects can devastate mature plants. Melons need delicate handling, they must be protected from contact with damp ground, and they need large quantities of space and nutrients to grow well.

The crucial factor in growing melons is soil warmth. Winter melons prefer a cool period for ripening, but American cantaloupe needs tropical conditions from as soon as possible after it is planted from seed or set out as a seedling, on through harvest. On the other hand, most varieties of cantaloupe ripen faster than winter melons. Seed germinates at 80°F, and vines grow fastest at daytime temperatures in the 80s and do well even at over 100°F. In spite of all their difficult cultural demands, melons, most gardeners agree, are well worth raising.

Melons are self-fertile and self-pollinating. Melons cross easily between varieties and all crossings remain fertile. If you want to save melon seeds, don't plant the crop near any other melons.

Light, fertile sandy or loamy soils that are rich in organic matter and have a pH of 6 to 8 are best for melons. They do poorly in heavy, poorly drained clays, but will stand light clay if it is well drained.

Melons are usually planted from seed, either directly in the plot, or in seedling trays or flats. They transplant poorly and are too delicate to ship as plants. In ordering seed, the most important consideration is your growing season. Count the days with proper growing temperature in your area and choose varieties on this basis.

The melon patch should be in a sunny spot, protected from winds, with good drainage and good air circulation. A southern slope or south-facing hill is ideal. Avoid spots with high water tables. Break up compacted soil and double-dig if possible.

Melons are usually planted in hills 3 to 5 feet apart in beds at least 5 feet wide. This method allows growing areas to be prepared by digging a pit the size of a bushel basket, working in manure, and heaping up soil over the dug-out area. Set six to eight seeds or three or four plants in each hill. Cover seeds with ½ inch of fine soil. As they develop, thin seedlings to three per hill.

Melons require heavy watering, especially when bearing flowers and small fruits, but they use less water when ripening. In extremely dry areas, a clay flower pot sunk into the melon hill and filled with water each day, will provide a constant water supply. In other areas, water for about half an hour, or until thoroughly soaked, in weeks in which no rain has fallen. A mulch will conserve water, but should not be applied until the ground is thoroughly warm.

Since melons are heavy feeders, applying manure to a whole melon patch in the fall or adding rotted manure to hills in the spring two weeks before planting is beneficial.

Feed plants with fish or seaweed emulsion or compost "tea" on planting, at time of fruit set, and about two weeks after fruit set. An excess of nitrogen can produce soft, odd-shaped fruit. High levels of potassium and phosphorus are, however, necessary, and the trace elements magnesium and boron, elements lacking in many soils, are essential to melons. Melon patches should be shifted in the garden every year for pest control and soil recovery.

The chief pests of melons are melon aphids, squash bugs, and striped and spotted cucumber beetles. Cucumber beetles spread fusarium wilt fungus, while aphids spread cucumber mosaic.

Handle melons, especially muskmelons, carefully at all stages of ripeness. When the vines go beyond the edges of a bed, lift the fruit gingerly and set it back in the bed center, rather than flipping it. Heap mulch under fruit to protect it from rot. Do not allow the mulch or ground under fruit to remain damp.

The first muskmelons begin to ripen 35 to 45 days after pollination, depending on the variety. Winter melons take somewhat longer. Each vine produces three or four fruits, although there are sometimes more in the true cantaloupes and in muskmelons. Ripeness is evidenced by the odor which permeates the garden where true cantaloupes or most varieties of muskmelons are grown. In some varieties, the skin color of muskmelons becomes yellow behind the netting. The best test for ripeness is to examine the stem for cracks. If the stem separates when pushed gently with the thumb, the fruit is ripe.

Some winter melons will not show stem cracks when ripe. Honeydew fruit must be cut from the vine when the blossom end is slightly springy. At this stage the fruit has reached normal size, has changed from green to white, and has lost much of its waxy look. A dull sound made when a honeydew or winter melon is thumped is also a sign of ripeness.

A good melon harvest will last two weeks. With a late and an early variety, a plot will supply melons for a month.

Peach and Nectarine

Peaches are grown in most of the major apple-producing regions though they can tolerate hotter weather and require less cold to break dormancy. Nectarines, the bald cousins of the peach, are nearly identical to peaches in the general appearance of the tree, growth habits, and bearing characteristics. An interesting horticultural oddity, it is possible for peach trees to grow from nectarine pits and vice versa and for a peach tree to sprout a limb bearing nectarines or the reverse. Their culture is so similar that all practices described here for peaches are also suitable for growing nectarines. Like peaches, there are white- and yellow-fleshed nectarines. Some varieties produce shocking pink, large-petaled flowers resembling a primrose. These include LORING, REDSKIN, RIO OSO GEM, and VELVET—all excellent landscaping trees. They have an attractive spreading shape with dark, glossy foliage and reddish brown bark.

Peaches, as a rule, are limited in their northern range by their inability to survive sustained temperatures below −20°F. Northern New England, the heart of the Great Plains, Alaska, and higher elevations in mountainous regions are unsuited for peach growing. In these areas, the tree itself will usually survive if the first winter or two after planting is relatively mild, but the blossom buds will be frozen probably five years out of six so no crop will be harvested.

Most peach varieties are self-fruitful and will set full crops without another variety for pollination, except for J. H. HALE, JUNE ELBERTA, and HALBERTA.

Peaches will grow well on any deep, well-drained, loamy soil with a pH of 6 to 8. Excessively well-drained sandy soils can be a problem in dry years due to their low moisture-holding capacity. Before planting, the soil should be tested so deficiencies or imbalances in pH and nutrients can be corrected.

Most garden centers will have a few peach varieties available, but at a large mail-order nursery the selection may include up to 75 varieties. Choose one-year-old stock that is intermediate in size, about 4 to 5 feet tall and ½ inch in diameter. Avoid the very small seedlings, since they may be genetically weak and may never develop into strong, healthy trees.

The question of whether a north- or south-facing slope is best for peaches is complex. In colder areas, where the weather in the spring warms gradually, a southern slope is best since the trees will bloom relatively early and will have a longer growing season before autumn frosts. Where spring temperatures fluctuate widely, a northern exposure is best. Ideally, trees should bloom as early as possible to increase your growing season, but not so early as to risk a freeze.

Peaches require a chilling period, between 700 and 1,100 hours of temperatures below 45°F for most varieties. Breeding programs have produced peaches for warmer regions that need as little as 50 hours. In areas where there is insufficient chilling, even though blossom buds are preset, the tree will not bloom and will have very sparse foliage.

The best tree site is in full sun all day, where the somewhat sensitive buds and flowers will not be exposed to extremely cold temperatures, but also one where the buds don't break too early in the spring and chance freezing. Not too cold or hot, not soggy or bone dry, enough weight to the soil but not a raw material for ceramics; peaches require a site that is intermediate in all these conditions. Nectarines should not be planted in areas with very humid, muggy summers, as brown rot will likely ruin most, if not all, of the crop.

Like most tree fruits, the best time to plant peaches is very early spring except in southern areas where fall planting is recommended. Plant trees that are fully dormant or just beginning bud-swell when the ground is moist, but not wet. Space the holes 20 to 25 feet apart depending on the fertility of the soil.

Peaches require little phosphorus or potassium, but show dramatic responses to nitrogen applications. If you are using a concentrated nitrogen source, apply it in the spring when buds swell. Manure provides nitrogen but it is available at a lower rate over a longer period and should be applied in the fall after the trees are dormant. This allows rains and melting snow to leach the nutrients down to the root zone where they'll be available in the spring. Wood ashes may be spread with the manure.

Peach trees need 1 ounce of nitrogen annually for every year of tree age up to 12 years, when the requirement for this nutrient levels off. This ounce may be supplied by 10 to 12 pounds of cow manure, 4 pounds of poultry manure, 10 ounces of fishmeal, 8 ounces of dried blood, or 14 ounces of cottonseed meal. Nectarines require slightly more nitrogen than peaches.

Weed control is necessary for producing large, sweet peaches, especially when the fruit begins to ripen and can't compete for water. Mulch heavily to keep the area under the tree weed-free, or cultivate by hand or with a rotary tiller. Be careful not to penetrate deeper than an inch or two to avoid damage to the roots.

If allowed to ripen all the fruit it sets, the tree will yield baskets of small peaches that are little more than a pit with a thick skin, and the tree may not be fully hardened for winter. In mid-June, when the "June drop," the shedding of unpollinated or abnormal fruits, is nearly finished, pick off cherry-sized peaches until only one fruit remains on every 6 to 8 inches of branch. Distribute the load by dividing it between both sides of the limb. A well-thinned tree will yield much larger fruit.

If a dry spell should develop, fruit will not size well and may become mealy, as the tree will withdraw the fruit's water for its own needs. Watering every other day with 25 gallons per tree will help until the rains return.

Train trees to an open-center form. Peach limbs need exposure to sunlight to remain healthy and will die if shaded for a season or two. Cut back the leader or main stem of the newly planted tree to

24 to 30 inches above the ground in order to encourage branching. Trim off any weak, dead, or broken branches that remain on the trunk. At this point the tree will look like no more than a stick, but it will very quickly leaf out and you can begin selecting the scaffold limbs. Let a foot or so of growth push out, then in the middle of the first summer, carefully select three or four shoots spaced evenly around the tree and cut off the rest. If there are insufficient shoots to make the scaffolds, fill in the empty spots the next summer. Don't remove any growth from those chosen shoots, allowing them to grow vigorously all summer.

Pruning in following years can be done anytime during the winter in warmer areas, and to reduce possible winter damage, in late winter or early spring farther north. Though the tree may set a few peaches the year after planting, concentrate on developing the open-center shape and a strong framework to bear later crops. Since peaches bear fruit on the previous season's wood, develop as much bearing surface as possible from which fruit-bearing shoots can grow. Leave some shoots in the center of the tree, but head them back to 12 to 15 inches. Remove any vertical growth or cut it back to an outward-growing shoot, since the idea is to keep the tree low, 8 feet but spreading.

The peach tree and lesser peach tree borers are probably the most persistent insect pests, while Japanese beetles, oriental fruit moths, and plum curculios may be occasional local problems. The adult female borer is a 1-inch clear-winged moth with a metallic blue body banded with one orange stripe, active from June through September with peak egg-laying in August. Eggs are deposited in crevices on the bark on the lower trunk and in the soil immediately around the trunk, and the larvae tunnel into the trunk. They will reveal their presence with small bits of sawdust mixed with their droppings around a hole in the trunk near the ground. A heavy infestation can seriously weaken or kill a tree. The lesser peach tree borer attacks the scaffold limbs, though seldom causes serious damage. The best medicine against borers is preventive: keeping your trees healthy and avoiding damage to the trunk where pieces of bark are torn away (favored egg-laying sites). Moth crystals around the trunk may act as a repellent, and a band of sticky Tanglefoot around the trunk down to the soil line will gum up any moth attempting to lay eggs.

As the fruit is sweetening just prior to harvest, watch for Japanese beetles and destroy them by handpicking and placing in soapy water. Oriental fruit moths will lay eggs in the growing shoot tips early in the season, causing them to wilt, while the later generation attacks the fruit, causing it to drop prematurely. If these become a problem, *Bacillus thuringiensis* will help.

Brown rot fungus heads the list of disease problems. Brown rot infections on twigs, flowers, and/or fruit must be removed as soon as they are noticed. "Mummies," dried infected fruits that remain on the tree over the winter, must be removed before the buds swell, and a light cultivation under the trees just before bloom will destroy any overwintering infections. Sanitation is critical in brown rot control, though powdered sulfur applied to the trees as a dust or spray will retard the growth of the fungus. To prevent bacterial leaf spot, plant varieties that are not susceptible to this disease. Cytospora or Valsa canker is a fungal infection of the wood which blocks the tree's circulatory system, causing sticky brown sap to ooze from the bark. Delay pruning until bud-swell so the cut surfaces can heal, and avoid breaking the bark. When canker is found, prune it out.

Nectarines are more susceptible than peaches to brown rot and, possibly due to their lack of fuzz, to damage from plum curculios and thrips.

As a peach begins the last phase of ripening it rapidly expands, and the size may increase 50 percent in the last three weeks. When it's first ready for picking, the flesh at the end away from the stem will give slightly to thumb pressure. This is considered firm-ripe. Peaches at firm-ripe will store in a refrigerator for two weeks and, brought out, will ripen to excellent flavor and texture at room temperature.

A peach will continue to grow and sweeten the longer it is left on the tree until it is tree-ripe, when the flesh near the stem end yields to thumb pressure. These can be held in the refrigerator, at best, only a few days. The flavor and melting quality of a tree-ripe peach is a real delicacy and worth the extra few day's wait.

To pick a peach without bruising it or damaging the tree, cup it in your hand and lift with a slight twist. The short stem will separate cleanly rather than tearing as it would if you were to pull it straight off.

Pear

The pear is first cousin to the apple, quince, and other pome fruits. Compared to apples, pears are harder to grow and more variable in quality, size, and shape. Among temperate tree fruits, they remain perennially in second place to the apple. Though they may be the apple's poor relation, pears grow well in approximately the same temperate climate where apples thrive, and in general are nearly as hardy.

Pear trees grow more upright than apples, and sometimes even taller, though standard sizes can usually be maintained at about 20 feet tall. Average size is that of an elongated apple, with some varieties, like SECKEL, being half the average size. The characteristic color is green ripening to yellow, though some varieties are red, and the yellows are often russeted and blushed with red or pink.

Dwarfing of pears has not progressed as rapidly as it has for apples, but commercial nurseries now sell many varieties dwarfed onto quince rootstocks, with an OLD HOME pear interstem for graft compatibility. These trees grow to only 10 feet and may need staking. A standard tree can be expected to produce 5, 10, or more bushels of fruit in a year; a dwarf, 1½ bushels.

Most pear trees need cross-pollination for good fruit set, and although there are exceptions, it is best to plant two varieties, or in some cases, three.

Pears, more than apples, prefer a heavier loam, but will grow in any well-drained soil.

Follow the general rules for planting described in Chapter 2. Set in the tree at approximately the same depth as it grew in the nursery bed, with the graft bulge above ground level. Then cut back the top so it roughly equals the amount of root. A yearly mulch of straw, hay, or light strawy manure with a cup of bone meal and several cups of wood ashes should suffice. Do not feed pear trees too much soluble nitrogen, because the lush growth such feeding can cause is especially susceptible to fireblight. Plant dwarf varieties 12 feet apart, standard varieties, 16 to 20 feet apart.

Prune pears as little as possible. They seldom respond well to corrective pruning and it may, especially in the summer, invite the lush growth that encourages fireblight. A pear tree's upright growth habit makes it necessary to head back side branches that want to overtake the central leader, but head back only cautiously and in the dormant season since the cuts also may induce lush growth. It's better to train branches out with spreaders or weights, or wait for a heavy crop of pears to bow the branches over. Prune off and burn branches affected with fireblight.

Fireblight, a bacterial disease, is the worst enemy of pears, and there is no cure. The best program to follow is to plant resistant varieties and avoid heavy fertilization. Pear psylla are the worst insect pests, having become resistant to many sprays. A late spray of water and dormant oil has been found to be effective in orchards where other psylla predators are active. But this spray must be timed exactly; you need a trained entomologist to predict just when the female psylla are going to emerge for egg-laying. Pear slugs are another pest, soft-bodied, dark green worms that feed on leaves.

The harvest season for pears generally runs from early August through September. For highest quality, pick pears by hand about two weeks before dead ripe and allow them to ripen slowly in a cool room. This practice eliminates some of the graininess that is characteristic of many tree-ripened pears. When it is green and firm, lift the fruit and the pedicel should separate from the twig.

To store for up to three months, keep pears at temperatures just above freezing in their preripened state, then thaw and keep at room temperature for a few days until ripe. Winter pears, which are still hard and inedible when they fall from the tree, can be put in any common storage area where they will soften during the winter and can be canned or used for other processed foods.

Plum

Plums fall generally into three groups: American, European, and Japanese. The pacific plum, beach plum, and the western sand cherry are native to North America and make excellent preserves and syrups, but are usually too tart to eat fresh.

Plums for eating, canning, or freezing are produced by two species — the European and Japanese plums — that have both been the subjects of much breeding work, resulting in many varieties. The DAMSON plum, a variety of a third species, should be considered as a special case; though its fruit is too tart to eat fresh, DAMSON preserves are of a quality that is in a class apart.

To clear up any confusion between prunes and plums: A prune is a dried European plum of certain varieties, which has a very high sugar content. It can be dried whole without fermentation around the pit. A plum that is split and dried with the pit removed is a dried plum, not a prune. All prunes are plums; not all plums are prunes.

A mature Japanese plum will produce an average of 1 to 1½ bushels of fruit. European types will yield twice that amount.

Plums do well in any deep, well-drained soil, though they will tolerate heavier soils better than most tree fruits. European plums root deeply and can penetrate a slight hardpan. For other considerations in choosing a site, see Peach.

Pollination in the plum world is a complex operation and bears careful consideration when choosing varieties. To start with, most plums require a second variety for cross-pollination, though a few, mostly European, types are self-fruitful. Some plums have sterile pollen, making a third variety necessary if all trees are to bear. For pollination to occur, the bloom periods of the varieties must, obviously, overlap. This is the main problem with relying on late-blooming Japanese types or vice versa.

As a rule European plums are similar in hardiness to apples and slightly hardier than the Japanese type, though some Japanese hybrids are quite hardy. They also make a much larger, more upright tree that blooms later, often avoiding damage from spring frosts. Japanese types with their often severely spreading habit can be a striking landscape addition, especially when the branches are covered with small white blossoms in spring. Chilling requirements are similar to those of peaches.

Plant a plum much as you would a peach, leaving a space 18 to 24 feet in diameter for Europeans, 14 to 16 feet for Japanese, and 10 to 14 feet for the native species, which tend to be large bushes rather than trees. It's important that the area around the tree (1 to 2 feet at planting, increasing to 4 to 6 feet) be kept weed- and grass-free to avoid competition for nutrients and water with the tree's subsurface feeder roots. Mulch is a good idea. A good soaking and a tree guard or wire mesh to stymie the mice are the final planting chores.

Requiring less fertilizer than peaches, a plum tree can be kept adequately fed with annual applications of barnyard manure 1 to 2 inches deep under the dripline. Spread it in late winter to allow the nutrients to begin leaching into the root zone where they will be largely consumed by late summer, when the tree must begin hardening for winter. A bucket of wood ashes, spread with the manure, will supply important mineral nutrients.

As with peaches, if you want to pick full-sized fruits rather than thousands of marbles, plums should be thinned after the "June drop" to a 2-inch spacing or so none are touching. A few varieties (such as SANTA ROSA and CLIMAX) will self-thin reasonably well.

Water is very critical to growing plums. Don't allow the trees to wilt or they may jettison the entire crop. Use the weeds near the tree as soil moisture indicators. When they're wilted early in the morning it's time to add 25 gallons of water per tree, twice a week.

Boron deficiencies may develop in some areas and can be recognized by dry, hard, sunken pockets in the flesh of the fruit. Spreading ½ pound of borax under the tree every third year will prevent this disorder.

One-year-old European trees are usually sold as unbranched whips and should be trained to the modified central leader shape, though they will form a sound tree capable of bearing heavy crops with little help from the pruner. Japanese plum seedlings are usually branched and with their spreading habit work best in the open-center shape with four or five main scaffolds. Prune trees lightly for the first few years; heavy pruning will delay fruiting and promote more vegetative growth that must later be removed. Japanese plums begin bearing in three to four years, the Europeans a year or two later.

Plums bear on one-year-old shoots and spurs on older wood. The object of pruning is to maintain seasonal growth of the one-year wood and to keep the center of the tree open to sunlight, without which the spurs will die. Plums, as a rule, are pruned more lightly than most fruits, except cherries. To reduce possible winter damage, prune in the early spring before the buds swell.

Birds, borers, plum curculios, and scale all prey on plum trees. They are also affected by brown rot. But the biggest enemy of the plum tree is black knot, a fungus known by the characteristic black, gnarled swellings it produces on affected shoots and limbs. Although black knot doesn't affect the fruit, if left unchecked it can girdle and kill major limbs, eventually killing the tree. Black knot can be eliminated over the winter, when its dark color makes it obvious. Cut the cankers out, removing 6 inches of unaffected wood below the cut to prevent reappearance, and burn the prunings.

Plums are green until just before ripening when the green gives way to red, purple, or yellow depending on the variety. Picked when firm-ripe, when the flesh is beginning to soften to thumb pressure, a plum can be held several weeks at 35°F, slightly cooler than a household refrigerator. But if you're going to the trouble of growing them, restrain your urge to harvest, give the plums another week, and pick them when tree-ripe and sweeter and more fragrant than possible otherwise. A tree-ripened plum, like a tree-ripened peach, is a rare treat that's only available if you grow it yourself, since they don't ship well.

Raspberry

The raspberry is a bramble native to the Orient. There are 200 species of raspberries native to eastern Asia, while Europe has only one and North America, three.

The raspberry bush usually grows to a height of 5 to 8 feet. Red raspberries have erect canes, while black raspberries have canes that arch or trail.

The berries or caps are shaped like thimbles, and range in size from ⅜ to 1 inch. When the fruit is picked it pulls away from a white cylinder which remains attached to the plant. The raspberry's hollowness distinguishes it from the blackberry, which it otherwise closely resembles.

Raspberries can tolerate both light and heavy soils, but cannot tolerate wetness. Soil should be well drained. A pH of 5.5 to 7 is adequate, but 6 is best.

Raspberry plants are available in a choice of varieties from many nursery catalogs. Because of the prevalence of viral infections in raspberries, buy virus-free stock even though it is more expensive.

Raspberries can be maintained in hedgerow plantings or as individual clumps. They are most often tied and trained to trellises, the designs of which vary with the grower. The simplest method is to plant raspberries in rows with a single overhead wire 3 feet high and set between posts. Wrap the canes around the wire or tie them to it. Cultivate the soil a whole year ahead of time, if possible.

Plants received from a nursery have a short length of cane above the roots and crown. Spread the roots out in the planting hole with the crown just barely under the soil surface. Press soil down firmly around the plant. New shoots will grow from the crown. Once they start growing well enough to mark the row, cut off the short length of old cane, even if it is sending out buds. It's new growth you want to encourage. Those first new canes will probably not produce fruit, but they will get the plant well established so that the next season's shoots will produce the first crop the year after planting. Raspberries are biennial. Shoots bear in their second year, then die. Space the plants 2 to 3 feet apart. Eventually they will fill in the row. Make rows 6 or more feet apart to leave ample space for cultivating or for mowing a grass walkway between rows.

Raspberry plants need a lot of water in the period just before late summer. In August and September, however, the amount of water should be decreased; too much water this late in their growth period may delay maturity of developing wood. This immaturity could lead to winterkill.

Add mulch or compost in May or early June to prevent the roots from drying out during fruit development. Mulching may make irrigation unnecessary, but if conditions are very dry, water from 1 to 2 inches a week.

Mulch in the fall as well, using grass clippings, leaves, straw, or manure. Raspberries like nitrogen, used sparingly. Phosphate rock is recommended every four years.

Different varieties require different pruning techniques. In addition, the red and yellow varieties are handled differently depending on whether they bear in summer only, or bear in both summer and fall.

Blacks and purples: Tip prune to prevent the canes from rooting new plants and to increase the crop the following year. Keep in mind that blacks bear on canes produced the previous year. After picking, cut out those canes at ground level, giving the new growth more room and sunlight. When the rapidly growing current season's canes reach 2½ to 3 feet tall, tip them to encourage laterals to develop. This, in turn, will increase the bearing wood/cane for the following year.

Once tipped, the laterals may grow 2 to 5 feet, depending on the vigor of the particular plant. If the laterals seem determined to tiproot, nip off a few inches. Otherwise, no pruning is necessary after the tipping. In the spring following tipping, and before growth begins, cut back the laterals to 10 inches. On less vigorous canes, cut back the laterals to only 6 to 8 inches. It's on these stubs that flowers, and later berries, will be borne.

If you have allowed some canes to root to fill in gaps in the rows, cut the mother cane loose from the new plant or leave it to produce berries low on the hedgerow. After berry harvest, cut out all the old fruiting canes. Thin the new canes, which will fruit the following year, to about four of the healthiest per foot of row if the row is maintained 2 feet wide. If the row is very narrow, thin more vigorously to one cane every 6 inches. Trial and error will reveal how many vigorous canes your particular soil can support.

Red and yellow summer bearers: Immediately after berry harvest, cut out all the old bearing canes to make room for the new canes growing amidst them. Thin the new canes as described above. Since reds and yellows spread by suckering, they are much harder to keep within the row bounds than are blacks. Despite all efforts they usually grow so thick by the end of five years that it is better to tear up the row and start again someplace else. Figure high production and big berries from a row for only that long, although by vigorously digging out extra suckers and old decayed roots, it is possible to keep a bed or row going until virus disease renders it unproductive. In spring pruning thin out the canes leaving only the heaviest spaced 6 inches apart, and cut them back to about 40 inches tall, being sure to cut off all winterkilled ends. Less vigorous canes should be headed at 36 inches.

Red and yellow summer and fall bearers: After summer harvest, cut out the old canes that have fruited. The new canes coming on will fruit in the fall, and then again the following spring. Thin out the weak ones nonetheless. After the fall crop has been picked (or more than likely, after frost has stopped the fall production), there are two possible ways to proceed. The first is to prune off the bearing tops of the canes, leaving about 3 feet of cane, which will grow out and produce a summer crop the next year. The second method avoids having to cut out old canes in the heat of summer; cut the whole hedge right down to the ground. Do not use a rotary mower for this chore. It will spread virus, if any is present, and infect the whole bed. A scythe or weed chopper is best. Remove and burn cut canes. There will be no summer crop the next year, but the new canes that grow next spring will produce a better crop without the competition of the summer-bearing canes.

Red raspberries are more disease-resistant than black and purple varieties. The black sap or juice beetle, often called the picnic bug, is the worst insect pest of raspberries. Viruses and anthracnose fungus present the most difficult disease problems. The best defense is to cut out old canes promptly after they have fruited. Moving healthy young plants to start a new row on a regular schedule (every five years) is also helpful.

Blacks are not as hardy as reds, but no raspberries need winter protection. To avoid winterkill do not fertilize or water in the fall, which would encourage lush growth into winter.

Harvest raspberries gently with thumb and forefinger. Place the berries in a shallow bowl to avoid crushing and bruising them. Keep them out of the sun. Pick every other day during the ripening season. Washing raspberries tends to make them lose texture, and is not necessary if they are grown without toxic chemicals.

Strawberry

Wild species of strawberry, native to both Europe and America, are small, extremely delicious and aromatic fruits that grow in woodland environments. In flavor and texture, they are more than the equal of cultivated strawberries. The large domestic berries developed from them are now grown throughout temperate regions. There is, for instance, at least one strawberry variety adapted to every state of the United States. Many varieties are, in fact, so regionalized that they do not perform well elsewhere.

Most varieties propagate themselves by runnering. The mother plant, soon after producing fruit, sends out runners that tiproot about a foot away. These new plants in turn send out their own runners. If left unchecked, plants will soon form a dense mat or bed.

In the first bearing year, young plants will bear heavily, as will their first runner plants, which root early in the summer. The next year, the runners that rooted later will bear heavily, the old mother plants will bear only a few small berries, and the patch will become so crowded that the new plants can find little space to root and grow. A partial solution is to continually dig out old plants. Better still is to rotate the strawberry patch every other year or every third year.

If an old bed is to be held over for a second bearing year, renew it mechanically and it will bear reasonably well again. Immediately after harvest, mow off the tops with a lawn mower. Then use a tiller to cultivate strips through the bed, leaving rows of plants about a foot wide. Cultivate the old plants, leaving the current season's well-rooted runners to bear the next year's crop.

Plants come from the nursery tightly wrapped in plastic. Store them in the refrigerator until ready to plant. At planting time, place them unwrapped in a bucket of water so the roots won't dry out. Make a little cone of soil with your fist and set the plant on the cone. Fan out the roots over the cone, pull in dirt, pat it down firmly, and water to settle the earth around the roots. It's important that the crown of the plant be right at the soil surface, neither so high that roots are exposed, nor so low that it is buried. If the roots are very thick and long, prune them to about 5 inches long with scissors before planting.

Spacing rows, like training runners, can be done in a number of ways. The way rows are spaced, however, depends on the method used to train runners. You can plant rows 4 to 5 feet apart and plants 2 feet apart in the row, and by allowing only four runners to set around each mother plant, have two well-defined rows producing very large berries the next year. Or you can allow the runners to form a matted bed.

Everbearing types cannot be treated as above, since they produce berries a second time in the fall. Everbearers are satisfactory only in those few climates where cool weather in late summer prevails.

Plants are set in the spring in the north, for the next year's crop. Farther south, plants can be set out in fall, for next spring's crop. To extend the harvest season, use a combination of early, midseason, and late varieties.

After the plants start growing, cultivate them once or twice to destroy early weeds, then mulch in June. Strawberries are shallow-rooted but need ample moisture. Keep the mulch fairly shallow immediately around the crown. Thick mulch may increase berry rot. Once the ground has frozen apply clean straw or similar clean material over the berry plants to inhibit frost heaving and possible plant damage. Wait until after freezeup, so mice do not take up winter quarters under the straw. When warm weather returns pull the straw off the plants into the row spaces or along the sides of the bed, where it will cushion the berries and keep them clean.

If the first mulch put around the plants (in the first year when no berries are harvested) is barn manure, the plants will get all the nutrition they need in a normal soil. Even straw alone is useful. Overfertilizing berries increases yields at the expense of flavor. That's also true of irrigating. In areas of adequate rainfall, mulch will provide adequate moisture for a backyard crop except in drought years.

Slugs and birds present the most serious problems. In rotated backyard patches, insect pests are common but not unusually troublesome. Buy disease-free stock to avoid verticillium wilt and red stele.

Strawberries ripen about a month after blossoming. Pick by pinching the stem between the thumbnail and forefinger; avoid pulling on the berry itself. Strawberries do not keep well for more than a day or two. The sooner eaten after picking, the better.

Rodale Press, Inc., publishes RODALE'S ORGANIC GARDENING®,
the all-time favorite gardening magazine.
For information on how to order your subscription,
write to RODALE'S ORGANIC GARDENING®, Emmaus, PA 18098.